Global Perspectives on Health Geography

Series Editor
Valorie Crooks, Department of Geography
Simon Fraser University
Burnaby, BC, Canada

Global Perspectives on Health Geography showcases cutting-edge health geography research that addresses pressing, contemporary aspects of the health-place interface. The bi-directional influence between health and place has been acknowledged for centuries, and understanding traditional and contemporary aspects of this connection is at the core of the discipline of health geography. Health geographers, for example, have: shown the complex ways in which places influence and directly impact our health; documented how and why we seek specific spaces to improve our wellbeing; and revealed how policies and practices across multiple scales affect health care delivery and receipt.

The series publishes a comprehensive portfolio of monographs and edited volumes that document the latest research in this important discipline. Proposals are accepted across a broad and ever-developing swath of topics as diverse as the discipline of health geography itself, including transnational health mobilities, experiential accounts of health and wellbeing, global-local health policies and practices, mHealth, environmental health (in)equity, theoretical approaches, and emerging spatial technologies as they relate to health and health services. Volumes in this series draw forth new methods, ways of thinking, and approaches to examining spatial and place-based aspects of health and health care across scales. They also weave together connections between health geography and other health and social science disciplines, and in doing so highlight the importance of spatial thinking.

Dr. Valorie Crooks (Simon Fraser University, crooks@sfu.ca) is the Series Editor of Global Perspectives on Health Geography. An author/editor questionnaire and book proposal form can be obtained from Publishing Editor Zachary Romano (zachary.romano@springer.com).

Joseph Asumah Braimah
Elijah Bisung • Vincent Kuuire
Editors

Health Geography in Sub-Saharan Africa

Development-Health Nexus

Springer

Editors
Joseph Asumah Braimah
University of Toronto Scarborough
Scarborough, ON, Canada

Elijah Bisung
Queen's University
Kingston, ON, Canada

Vincent Kuuire
University of Toronto – Mississauga
Mississauga, ON, Canada

ISSN 2522-8005 ISSN 2522-8013 (electronic)
Global Perspectives on Health Geography
ISBN 978-3-031-37564-4 ISBN 978-3-031-37565-1 (eBook)
https://doi.org/10.1007/978-3-031-37565-1

This Springer imprint is published by the registered company Springer Nature Switzerland AG
The registered company address is: Gewerbestrasse 11, 6330 Cham, Switzerland

Contents

Contributors

Ellis A. Adams, Ph.D., is an assistant professor of Geography and Environmental Policy at the Keough School of Global Affairs, University of Notre Dame. He is affiliated with Notre Dame's Environmental Change Initiative and the Eck Institute for Global Health. Trained as a human environmental geographer with expertise bridging the natural and social sciences, he is broadly interested in nature-society relations. His current work examines the social, political, institutional, and governance dimensions of environmental and natural resources, particularly water, and he conducts research in Ghana, Malawi, Kenya, Uganda, and the United States.

Benjamin D. Agbemor is a doctoral candidate in the Department of Geography at Kent State University in the USA. He received a Master's degree from the University of Leeds in the UK and a Bachelor's degree from the Kwame Nkrumah University of Science and Technology, Ghana. Benjamin has experience in water governance, water service monitoring, and household water provision. Benjamin has previously worked on water and sanitation projects in Ghana with public and private sector institutions. He is passionate about localizing the implementation of the UN Sustainable Development Goal six. He is also involved in research on household water insecurity, water supply alternatives, and environmental migration in urban Ghana.

Daniel Amoak is a doctoral candidate in the Department of Geography and Environment at Western University. Daniel is an SSHRC Doctoral Awards recipient. His research broadly focuses on sustainable food systems and rural development. He also has a research interest in climate change resilience, population health, and infectious diseases in the Global South.

Florence Wullo Anfaara is a doctoral and Vanier Canada scholar in Gender Studies and Transitional Justice at Western University. Her Ph.D. work seeks to document and archive how Liberian women's groups mobilized at the grassroots level to help bring an end to the 2014 Ebola crisis. She researches topic areas in gender, peace, and conflict studies, infectious and communicable diseases, social determinants of health and wellbeing, and African feminist theory.

Eunice Annan-Aggrey is a doctoral candidate in the Department of Geography and Environment at the University of Western Ontario in London, Ontario, Canada. Her research interests include international development, the Sustainable Development Goals, local governance, gender equality, poverty reduction, and environmental sustainability.

Roger Antabe, Ph.D., is an Assistant Professor in the Department of Health and Society at the University of Toronto Scarborough. He is a health geographer whose research interests span both the Global South and North (specifically, sub-Saharan Africa (SSA) and Canada). In SSA, his research focuses on environmental exposures, population health inequalities, healthcare access, and utilization of health services especially among marginalized and structurally exposed populations. His research focus in Canada is on the poor health outcomes of racialized populations and immigrants at the nexus of behavioral and structural risk factors.

Godwin Arku, Ph.D., is a professor in the Department of Geography at the University of Western Ontario, London Ontario, Canada. His research interests span the 'urban' and 'economic' sub-division of human geography, especially the transformation of urban systems in a changing global environment. He is also interested in issues of third world development, especially in Africa. Professor Arku is also the Editor in Chief of the *African Geographical Review*.

Kilian Atuoye, Ph.D., is an assistant professor in the Department of Global Development Studies at Queen's University, Canada. He is a health geographer with research interest focused primarily on health equity. His work spans environment and health, social epidemiology, and healthcare access, and embraces global, community, and individual health perspectives. He is particularly interested in the environmental and social production of human health and health inequalities, with a focus on determinants, impacts, and the policy environment of health and healthcare.

Elmond Bandauko is a doctoral candidate and SSHRC Vanier Scholar in the Department of Geography and Environment at the University of Western Ontario in London Ontario, Canada. His research interests include urban transformation in African cities (gated communities and new cities), smart cities, gender and urban development, urban policy, housing struggles of the urban poor, and urban informality in cities of the developing world.

Elijah Bisung, Ph.D., is an assistant professor in the School of Kinesiology and Health Studies, Queen's University, Canada. He is a health geographer with research interest that focuses broadly on social and environmental production of health and wellbeing.

Joseph Asumah Braimah, Ph.D., is a UTSC Inclusive Excellence Postdoctoral Fellow with the Department of Health and Society, University of Toronto Scarborough, Canada. His research focuses primarily on the health and wellbeing of marginalized and structurally exposed populations in the Global South and the Global North, drawing on multiple methodologies and theoretical perspectives. His work largely revolves around aging and health, social determinants of health, food security, and community-based participatory research.

Laifolo Dakishoni is the research coordinator of the not-for-profit Soils, Food and Healthy Communities (SFHC) organization in Malawi. SFHC carries out participatory research with smallholder farmers on agroecological and community-based approaches to address food security, nutrition, and social equity. He has co-authored several articles in this area.

Ebenezer Dassah, Ph.D., is a lecturer with the Department of Global and International Health at the Kwame Nkrumah University of Science and Technology (KNUST), Ghana. He is researcher with research interests that broadly focuses on global health and disability. His research particularly involves conducting policy and practice relevant health research to address health inequities among marginalized populations (i.e., persons with disabilities, older adults, and children) globally.

Mark A. Deka, Ph.D., is an ORISE Fellow with the Centers for Disease Control and Prevention in Atlanta, Georgia. He is a medical geographer with research interests that span topics like vector-borne and zoonotic diseases, ecological niche modeling, and spatial epidemiology.

Jenna Dixon, Ph.D., is a research associate in the Faculty of Health and Social Development at the University of British Columbia, Okanagan. She brings her training as a health geographer to research issues in global health, with a particular interest in health equity science and knowledge mobilization.

Obinna C. Ezeagbor is a Master of Arts degree student in the Department of Sociology, Memorial University, Newfoundland.

Catherine Hickey has a Master's degree in Geography, and works as a project coordinator at Western University, Canada. Her work has included projects focused on agroecology, food security, gender, climate change adaptation, policy, and maternal and child health.

Hilary Hungerford, Ph.D., is an associate professor in the Department of Earth Science, Utah Valley University. She is a human geographer with a research interest in human-environment interactions. She also works on sustainability issues, both in West Africa and in Utah.

Joseph Kangmennaang, Ph.D., is a Queen's National Scholar in Black Health and Social change, and his primary research examines how the places we live, work, worship, and play impact population health and wellbeing. His current research explores Black immigrants' experience in the United States and Canada toward understanding how Black immigrants' health and wellbeing are impacted by social, technological, and demographical changes. He employs social theories, participatory, and mixed method approaches to answer various questions and work with marginalized communities to promote their health and wellbeing. He is committed toward impactful and transformational research in global Black health.

Moses Kansanga, Ph.D., is an assistant professor of Geography and International Affairs in the Department of Geography in George Washington University. He is a health geographer, and his research spans three broad areas, namely, agriculture and sustainable food systems, natural resource development, and environment and health.

Yvonne Kasine is a doctoral candidate in Nursing and works as a teaching assistant in the Arthur Labatt Family School of Nursing and Research Assistant at the Center for Research and Innovation, the University of Western Ontario. Yvonne's Ph.D. research explores nurses and midwives' lived experiences of participating as mentees in an interprofessional clinical mentorship program in Rwanda. Yvonne has over 10 years of teaching experience in higher education institutions both in Rwanda and Canada. Yvonne has experience in Qualitative Research and Curriculum Development and Evaluation. Her research work has presented her with opportunities to work with incredible world-class researchers on both local and international health-related projects.

Mangani Katundu, Ph.D., is an associate professor of Food Security and Nutrition at the University of Malawi. His research interests include broad themes on sustainable agriculture, organic farming, conservation agriculture, and nutrition and food security with a primary focus in Southern Africa. He is a co-investigator of the Soils, Food and Healthy Communities and the Malawi Farmer-to-Farmer Agroecology project.

Rachel Bezner Kerr, Ph.D., is a professor in Development Sociology at Cornell University. She is a development sociologist with a background in environmental science, anthropology, and international development. She currently serves on a project team of the High Level Panel of Experts on Food Security and Nutrition of the United Nations and the Coordinating Lead Author for Chapter 5 on Food for the next Intergovernmental Panel on Climate Change report. She is also the director of the Community Food Systems minor in Cornell University, which provides engaged learning experiences for students with organizations working on sustainable agriculture and food justice issues.

Daniel Kpienbaareh, Ph.D., is an assistant professor of Geography at the College of Arts and Sciences at Illinois State University. He is a broadly trained geographer with an emphasis on human-environment interactions and their impacts on socio-ecological systems. He uses geospatial technologies to explore these interactions and employs social science theories to identify policy alternatives to environmental challenges.

Senanu Kwasi Kutor is a doctoral candidate in the Department of Geography and Environment at the University of Western Ontario, London, Canada. His research interests span transnationalism, geographies of wisdom, immigration and wellbeing, migration, immigrants' integration, and urban informality in cities of the developing world.

Vincent Kuuire, Ph.D., is a health geographer and currently an assistant professor in the Department of Geography, Geomatics and Environment at the University of Toronto – Mississauga. He also holds a Cross-appointment in the Social and Behavioural Health Sciences Division at the Dalla Lana School of Public Health, University of Toronto – St. George. Dr. Kuuire is the Tier II Canada Research Chair in Immigrant Wellbeing and Global Health, and his research interests and ongoing projects focus on the health and wellbeing of minoritized populations, global health, and health inequities.

Medjatu Kuyateh is an epidemiologist at Cabarrus Health Alliance working with the Performance Management and Quality Improvement Team. Her research background includes maternal and infant health disparities, communicable diseases in sub–Saharan Africa, and oral contraceptive use among immigrant and US-born women. Before joining Cabarrus Health Alliance, Medjatu worked as a consultant with Adam Smith International (ASI) in Freetown, Sierra Leone. Medjatu also led a team of volunteers delivering oral polio vaccines to children living in hard to reach communities and was among a group of first responders providing aid to survivors of a national disaster in Sierra Leone.

Isaac Luginaah, Ph.D., is a distinguished university professor in the Department of Geography and Environment at the University of Western Ontario, Canada, and a Fellow of the African Academy of Sciences. He was a Canada Research Chair in Health Geography (2007-2017). His broad area of research interest includes environmental exposure and population health.

Nasong Anthony Luginaah is a student in the Master of Public Health program in the Schulich School of Medicine and Dentistry. He previously completed a Bachelor of Science specializing in Synthetic Biology. He has research interest in population and community health.

Esther Lupafya is a health and gender coordinator at Soils Food and Health Communities (SFHC) Organization. She is one of the facilitators for Curriculum training using drama in the community. She is a community nurse and holds an M.A. in Social Development and Health obtained from Queen Margaret University, Scotland.

Niaz Morshed, Ph.D., is a Research Specialist for the Demography/GIS team under the Office of Data, Analytics, and Performance (DAP) of Texas Health and Human Service Commission. His research interests include human health and environment, racial disparity, quantitative methods, GIS, and big data analytics. He is also experienced with program-specific demographic and geo-spatial data analysis.

Marie Paul Nisingizwe is a doctoral candidate in the School of Population and Public Health at UBC, and her research focuses on access to hepatitis C testing and treatment in low- and middle-income countries. Marie Paul has seven years of experience in quantitative data analysis and health systems research. She previously worked with Clinton Health Access Initiative, Partners in Health, and the World Bank, where her work focused on health policy evaluation, infectious diseases, and maternal and child health. She is the recipient of the 2014 Harvard School of Public Health McGoldrick Fellowship and the 2020 World Bank Group Africa Fellowship.

Mark W. Rosenberg, Ph.D., was a professor of Geography in the Department of Geography and Planning and cross-appointed as a professor in the Department of Public Health Sciences at Queen's University in Kingston, Ontario, Canada. He was also the Tier 1 Canada Research Chair in Aging, Health and Development. He is now Professor Emeritus. He continues to do research on aging, health, health care, and health and the environment. Publications from his research can be found in the leading journals of geography, gerontology, social science, and medicine.

Yujiro Sano, Ph.D., is a postdoctoral fellow in the Department of Sociology and Anthropology at Nipissing University, North Bay, Ontario Canada. He is interested in health inequality and migrant and ethnic relations.

Lizzie Shumba has worked for over a decade for the Soils, Food, and Healthy Communities project in Ekwendeni, Malawi, and currently works with the Malawi Farmer-to-Farmer Agroecology project at Ekwendeni Hospital. She has a college diploma in nutrition from Natural Resources College, Malawi. She is passionate about improving nutrition using agroecological and participatory approaches.

Sarah L. Smiley, Ph.D., is a professor in the Department of Geography at Kent State University at Salem, USA. She is an urban geographer with research interests in water access, water insecurity, and development in Sub-Saharan Africa.

Eric Y. Tenkorang, Ph.D., is a professor of Sociology, cross-appointed to the Division of Community Health and Humanities at Memorial University. He is a Harry Frank Guggenheim Distinguished Scholar and a Member of the Royal Society of Canada (The College of New Scholars, Artists and Scientists). Dr. Tenkorang has served as a member of the Institute Advisory Board for Gender and Health of the Canadian Institute of Health Research (CIHR). He has broad research interests in population health, especially in limited-resource settings. This includes investigating the sexual and reproductive health of vulnerable and marginalized populations in sub-Saharan Africa. His most recent research has explored links between gender-based violence and health outcomes.

Germaine Tuyisenge, Ph.D., is an assistant professor of Sexual and Reproductive Health at the London School of Hygiene and Tropical Medicine. Germaine's research looks at access to sexual and reproductive health services among marginalized communities. She has conducted research in Rwanda and Canada and has worked on partnership programs in health education between the two countries, and currently between the UK and Ghana. She serves as the program director for the master's program in sexual and reproductive health policy and programming in low- and middle-income countries at the London School of Hygiene and Tropical Medicine.

Chapter 1
Health Geography in Sub-Saharan Africa

Joseph Asumah Braimah, Elijah Bisung, and Vincent Kuuire

Introduction

This edited collection aims to showcase sub-Saharan Africa's diverse health geography scholarship. In the context of the increasing need for evidence-based policy in the global south to tackle emerging health challenges (e.g., vector-borne diseases) which might otherwise be old problems, along with moves by universities to create "global classrooms" and internationalize teaching content, this collection is very timely. In terms of the development-health nexus, sub-Saharan Africa (SSA) can be characterized as a region of parallel health determinants and outcomes. On the one hand, the region is home to some of the highest incidences of poverty and low standards of living. In addition, poverty-related conditions (e.g., limited access to health care, food insecurity, famine and malnutrition) along with the high prevalence of many infectious diseases (e.g., HIV/AIDS, cholera) continue to afflict millions across the region. At the same time, the sub-continent is witnessing rapid economic growth and improved standards of living for sections of the population.

J. A. Braimah (✉)
Department of Health & Society, University of Toronto Scarborough, Toronto, ON, Canada
e-mail: j.braimah@utoronto.ca

E. Bisung
School of Kinesiology and Health Studies, Queen's University, Kingston, ON, Canada
e-mail: elijah.bisung@queensu.ca

V. Kuuire
Department of Geography, Geomatics & Environment, University of Toronto Mississauga, Toronto, ON, Canada

Social and Behavioral Health Sciences Division, Dalla Lana School of Public Health, University of Toronto – St. George, Toronto, ON, Canada
e-mail: vincent.kuuire@utoronto.ca

J. A. Braimah et al. (eds.), *Health Geography in Sub-Saharan Africa*, Global Perspectives on Health Geography, https://doi.org/10.1007/978-3-031-37565-1_1

The region is experiencing some of the most rapid urbanization rates, increasing the availability of specialized health services, technological advancements, growth of the middle class, and associated lifestyle changes—all under the broader umbrella of globalization. As a result, although infectious diseases continue to be the most important cause of mortality and morbidity, other important trends such as an increasing life expectancy and growing prevalence of non-communicable diseases (NCDs) are also becoming prominent (Defo, 2014; Holmes et al., 2010).

Amid the multiple and intersecting health determinants and associated health outcomes in SSA, concepts and frameworks for describing population health are essential for understanding and designing intervention programs to improve population health outcomes. For example, even though the region has a very young population, demographic projections show that the number of people above 60 years will quadruple by 2050. While life expectancy is increasing, the pace of fertility decline has been slowest in SSA than anywhere else in the world because of cultural, social, and economic constraints that impact women's reproductive health (Defo, 2014). These demographic changes come with demands for age-appropriate health services for all sections of the increasing population. With fragile health systems and lack of coherent policies, researchers argue that aging and demographic shifts in SSA countries will be uniquely different and require close attention (WHO, 2012; Kahn et al., 2007).

The unique changing demographic and disease profiles are also happening at a time when the effects of climate change are gaining prominence, and inequalities are accelerating—all with serious consequences for health disparities and social stability. For example, while economic growth in the past two decades has lifted millions out of poverty in SSA, income inequalities are a dominant feature of the economic growth in most countries, with 10 of the 19 most unequal countries in the world located in SSA (Odusola et al., 2017). This means that those who face the severest burden of disease and poverty are increasingly being left behind, which can have detrimental consequences on political stability, social cohesion, and general well-being (Berardi & Marzo, 2017; Stewart, 2011). Similarly, climate change is exerting pressure on old environmental challenges such as access to water and agriculture, which is in part leading to environmentally induced migration, food security, and illegal mining, with associated effects on health and well-being (Kuuire et al, 2021; Armah et al., 2016).

With the scale and changing socio-cultural dynamics of the challenges discussed in the preceding text, health geographers can uniquely advance knowledge on the drivers of individual and population patterns, distribution, and experiences of health and diseases in SSA. This collection provides a platform where both old but persistent problems, such as health care access disparities and the burden of infectious diseases, are discussed, while concurrently paying equal attention to emerging new challenges, such as NCDs and water-related stress. The collection also pays particular attention to how theoretical and methodological innovations in the sub-discipline can be applied to the SSA context. The collection consists of theoretically and empirically grounded contributions from a variety of disciplines—that draw on different epistemologies and methodologies—from different regions of the sub-continent (e.g., South, East, and West Africa), thereby creating a broader appeal for the book in other (sub)disciplines. As a collection that introduces new ways of thinking temporally and spatially about these topics in non-geography disciplines, it is our hope that it provides a platform for initiating the needed discussion on the complexity of issues impacting health in the region.

Geographies of Health in Sub-Saharan Africa

Health geography (or geography of health) emerged as a sub-discipline in human geography in the 1980s as a critique of medical geography, which was primarily concerned with the biology of diseases and the distribution of health services in line with positivist philosophies (Kuuire & Dassah, 2020; Moon, 2020). Health geography is concerned with understanding how space and place pattern the health and well-being of populations—based on a recognition of the complex "interaction between people and the environment" (Dummer, 2008: 1177). Since its inception, health geography has evolved considerably with the fundamental objective of interrogating social, economic, cultural, and political processes for health (Dummer, 2008). Within the sub-field of health geography, place is not only physical space, but include experiences influenced by social, economic, and political processes (Kuuire & Dassah, 2020). While debates exist regarding geographers' engagement with health issues and the naming of the sub-discipline, this section focuses on the nascent field of health geography in sub-Saharan Africa (SSA), particularly the topics dominating the budding literature in health geography in the sub-region. The evolution and debates surrounding the sub-discipline have been extensively discussed elsewhere (see Kearns, 1994; Luginaah, 2009; Rosenberg, 1998).

Over the past decades, SSA has witnessed an unprecedented explosion of cross-cutting empirical research on critical health-related issues. Concomitantly, researchers have been drawing on geographical perspectives and methods to explore health and well-being in SSA (Braimah & Rosenberg, 2021; Rishworth et al., 2016). Underlying this burgeoning research area is the role of place and space in influencing health outcomes. Research in health geography in SSA is particularly influential in addressing questions around disparities in health, gender-health nexus, and climate change effects, among others. As the sub-discipline gains prominence in the sub-region, not only are the topics broadening in scope and breadth in response to the complex and evolving health challenges, but also the methodological and theoretical approaches employed in conducting research here are expanding. Thus, new and innovative approaches, including geographic information systems, continue to permeate the sub-discipline. Furthermore, health geography scholarship in SSA, like other sub-regions, is not mutually exclusive but related to other cognate disciplines such as public health, sociology, and epidemiology. In the ensuing paragraphs, we provide insights on some of the dominant research themes in health geography in SSA, which include *health inequalities and health service use*, *marginalized and structurally exposed populations*, and *Water, Sanitation, and Health*.

Health inequalities and health service use is perhaps the largest thematic area of focus for health geographers in SSA. This is not surprising in the context of the region's complex demographics, disease profiles, and health systems (Oleribe et al., 2019). Health geographers' engagement under this research theme in the sub-region traditionally focused on the burden of diseases and the distribution of health services in line with medical geography. Quite recently, however, research under this thematic area has expanded considerably to include broader socioeconomic and environmental factors and their impacts on population health and well-being, such as health systems and health service use (Braimah et al., 2019; Woods et al., 2019), neighborhood environments and health (Amegbor et al., 2020), health financing (Amegbor et al., 2019; Kansanga et al., 2018), and food security (Atuoye & Luginaah, 2017; Braimah & Rosenberg, 2022). The primary objectives of research in this stream are to (1) understand contextual factors influencing health and well-being and (2) facilitate the optimum allocation of health resources, including health facilities.

Like the global north, research within the sub-discipline of health geography in SSA is also beginning to shift attention toward marginalized and structurally exposed populations such as older people, persons with disabilities, children, and women. Specifically, ongoing demographic shifts reflected in the rapid increase in the number and proportion of older adults in SSA have prompted research that explores health and social care for older adults (Braimah & Rosenberg, 2021; He et al., 2020; Kuuire et al., 2021; McQuaid et al., 2021). This research area is particularly crucial as national economies are not well positioned to accommodate this demographic shift. Additionally, there is a substantial amount of health geography research focused on maternal and child health in the sub-region (Sano et al., 2018; Tuyisenge, 2021).

The health geography literature is also characterized by research on Water, Sanitation, and Health (WASH). This is because research has consistently established a complex and positive relationship between access to safe water and improved sanitation and health (Bisung & Elliott, 2014; Dery et al., 2022). Specifically, geographic concepts and methods have been employed to examine the nexus between access to water and health by Anthonj (2021) and Bisung et al. (2015). The underlying assumption of research under this strand is that experiences or access to WASH differ in terms of geographic location, socioeconomic status, level of vulnerability and resilience to climate-related stress, cultural context, and the nature of the health system, among others (Bisung et al., 2015; Kangmennaang et al., 2020).

Indeed, health geography as a sub-discipline is well placed to tackle many health challenges faced by countries in the global south. Theoretical, conceptual, and methodological developments within the sub-discipline have opened new ways of thinking about and studying the complex determinants, patterns, and experiences of diseases and health care. As observed by a leading health geographer, with increasing spatiotemporal inequalities, health geographers in SSA ought to refocus their efforts on those individuals that are marginalized and structurally exposed (Rosenberg,

2017). The increase in the visibility of research by health geographers in terms of breadth and scope underscores the utility of the sub-discipline of health geography in providing intervention directions for addressing the increasingly complex health issues facing humanity in the global south.

The Nexus Between Health and Development

Health and development have been well documented to have a bidirectional relationship (Bisung et al., 2018; Luginaah et al., 2015). Thus, the health of a population depends on the level of development and vice versa. The World Health Organization defines health as "a state of complete physical, mental, and social well-being and not merely the absence of disease or infirmity (WHO Centre for Health Development (Kobe, Japan), 2004: 28)." On the other hand, development is a dynamic process that has evolved over the years. The United Nations Development Program conceptualizes development as an undertaking to achieve a higher quality of life. Sen conceptualizes development beyond this traditional focus on achieving better life to be an integrated process of increasing people's freedoms (Sen, 1988). Thus, development in its holistic view involves the removal of unfreedoms such as poverty, poor infrastructure, and systematic social deprivation, which are crucial to promoting health and well-being. Consequently, promoting health and development are the shared concern of health geographers.

Recognizing that health cannot be disassociated from development, all the ongoing United Nations Sustainable Development Goals (see Fig. 1.1), explicitly or

Fig. 1.1 The sustainable development goals. (*Source:* Gavi (Accessed 12/12/2022)) *Copyright*: **https://www.gavi.org/our-alliance/global-health-development/sustainable-development-goals**

implicitly, commit to promoting health and well-being. Therefore, it is not surprising that health remains a crucial indicator or measure of progress toward meeting these goals. It is important to state, however, that discourses on the development-health linkage often assume the standard of living as influencing health to the neglect of other equally important dimensions of development, including the quality of the institutions governing individuals, the environment people live in, as well as the political and economic systems, which have also been well established to define health and well-being (Bisung et al., 2018). While significant progress has been made in the context of health outcomes in SSA, wide variations occur in terms of geography and disease-type. These variations in health can be linked to spatiotemporal disparities in development in the given context. Thus, discourses on the development-health nexus ought to move beyond the traditional focus on the standard of living to focus on how socio-cultural and ecological factors shape health and well-being.

Overview of Sections

We have organized this book into four sections, each with two or three contributed chapters. The sections provide an organizational structure for the book and thematically place similar chapters together. The sections also capture some of the dominant themes within the sub-discipline of health geography in SSA. The empirical studies focus on many countries including Angola, Democratic Republic of Congo, Ghana, Rwanda, Malawi, Nigeria, Sierra Leone, and Tanzania.

The first section focuses on non-communicable diseases in Ghana. In Chap. 2, Kuuire and colleagues examine the influence of neighborhood inequalities and individual characteristics on non-communicable disease (NCD) multimorbidity in Ghana. The authors found that the neighborhood of residence is responsible for about 18% of the variance in NCD multimorbidity in Ghana, emphasizing the importance of place effects on the rise of NCDs in the sub-region. Ezeagbo and Tengkorang examine the determinants of hypertension among women in Ghana in Chap. 3. Recognizing that the rise in NCDs including hypertension poses serious challenges to Ghana's epidemiologic transition, the authors emphasize the need for behavioral change interventions that are gender sensitive and focus on changing the lifestyle of Ghanaian women. The final chapter in this section is a scoping review that synthesizes existing qualitative evidence on the strategies adopted by persons with mental illness to manage stress in Ghana, where psychiatric care is limited. Faith-based healing and prayers were the most common coping strategies identified in the review. Other strategies include seeking biomedical care, maintaining positive relationships, substance use, listening to music, and isolation. The review calls for coordinated mental care service provision and the need for increased research on mental illness in Ghana.

In the second section, we present studies which focus on maternal health. In the first study in this section (Chap. 5), Braimah and colleagues examine the connection between residing in functional community-based health and planning service (CHPS) zones and pregnant women's delivery choices in health facilities. The study interrogates the extent to which such decisions vary based on rural and urban residence. The positive impacts of functional CHPS zones on delivery choices in health facilities with maternal health experts lead the authors of the study to recommend expanding the CHPS initiative to cover remote locations where access to healthcare services is generally poorer. Anfaara and colleagues (Chap. 6) use data from the Demographic and Health Survey of the Democratic Republic of Congo (DRC) to examine the association between decision-making autonomy (i.e., women's autonomy) and exposure to family planning messages in mass media among postpartum married women. The authors emphasize the importance for stakeholders to pay attention to socio-cultural gendered norms in rural contexts. Based on the findings, they argue for the implementation of culturally sensitive community-level educational programs on gender equity. Chapter 7 takes a different approach, presenting reflections by three emerging scholars on navigating positionality in maternal health research in Rwanda. The authors discuss their roles as insiders to the communities they researched based on their familiarity, yet were considered outsiders by their participants since they were living and studying outside Rwanda for several years. The reflexive approaches adopted, and insights provide important examples for scholars conducting research in similar settings across Africa and beyond.

The third section deals with environment-health nexus in SSA. Recognizing that billions of people around the world are struggling to meet their daily WASH needs including large portions of SSA's population. Smiley and colleagues in Chap. 8 use case studies from the DRC, Ghana, Niger, and Uganda to identify the potential impact of COVID-19 on vulnerable and water-insecure populations. The chapter highlights why well-intentioned WASH interventions to control COVID-19 in the region failed to address the specific needs of poor, under-resourced, and vulnerable populations. Kansanga and colleagues (Chap. 9) use longitudinal data from the Malawi-Farmer-to-Farmer Agroecology (MAFFA) intervention to examine the impact of participatory agroecology on self-rated health. Because of concerns about the negative environmental and health impacts of industrial agriculture, agroecology has evolved as an alternative approach that draws upon local knowledge and resources to generate a cost-effective and environmentally sustainable farming system. Findings from the study show that households practicing agroecology are more likely to have improved health. Overall, the findings provide useful insights for policymakers to draw upon agroecology for the improvement of household nutrition and health toward the realization of the Sustainable Development Goals. Chapter 10 discusses gender-responsive urban development with examples and demonstrates the significance of mainstreaming gender into urban planning, housing, and service delivery (water and sanitation). The chapter argues that women in SSA cities are often marginalized in decision-making processes regarding the production of urban spaces, and they continue to endure the hardships associated with poor urban service delivery. Strategies to promote inclusive cities and prioritize women's health needs in urban areas are urgently needed.

The final section focuses on infectious diseases and discusses emerging and remerging infectious diseases (Ebola) and neglected tropical diseases (Tungiasis). The first study in this section by Kangmennang and colleagues uses a nationally representative household survey to explore the social determinants of non-participation in social life following the 2014 Ebola virus disease (EVD) outbreak in Sierra Leone. The study found that respondents who reported EVD exposure were more likely to report nonparticipation than those who were not exposed to EVD. Understanding the outbreak's non-EVD health impacts and pathways is important for rebuilding the health systems in affected countries. The 2020 COVID-19 pandemic has also reinforced the need to understand social outcomes associated with epidemics and pandemics. Finally in Chap. 12 of this section, Deka and colleagues investigate geographic and environmental factors associated with Tungiasis at multiple scales of analysis with ecological niche modeling (ENM) in East and Central Africa. The geography of Tungiasis is governed interchangeably by a variety of environmental factors. These include the density of livestock and human population densities, as well as low accessibility to urban areas, the dry season normalized difference vegetation index (NDVI), annual mean temperature, precipitation seasonality, precipitation of the wettest month, and cropland biomes. In the concluding chapter, Dixon and Luginaah contextualize how the different studies exemplify health geography and make a strong case about how these concrete examples signal the birth of a health geography of SSA. The chapter identifies complexities and nuances which found expression in the case studies in the book including individual, structural, environmental, and spatiotemporal factors that impact health outcomes and the understanding of conducting health research on the continent. Through the identification of these complexities, the concluding chapter shows the relationships between the various studies in the book.

References

Amegbor, P. M., Kuuire, V. Z., Bisung, E., & Braimah, J. A. (2019). Modern or traditional health care? Understanding the role of insurance in health-seeking behaviours among older Ghanaians. *Primary Health Care Research & Development, 20*(e71), 1–8. https://doi.org/10.1017/S1463423619000197

Amegbor, P. M., Braimah, J. A., Adjaye-Gbewonyo, D., Rosenberg, M. W., & Sabel, C. E. (2020). Effect of cognitive and structural social capital on depression among older adults in Ghana: A multilevel cross-sectional analysis. *Archives of Gerontology and Geriatrics, 89*, 104045.

Anthonj, C. (2021). Contextualizing linkages between water security and global health in Africa, Asia and Europe. Geography matters in research, policy and practice. *Water Security, 13*, 1–11. https://doi.org/10.1016/j.wasec.2021.100093

Armah, F. A., Boamah, S. A., Quansah, R., Obiri, S., & Luginaah, I. (2016). Working conditions of male and female artisanal and small-scale goldminers in Ghana: Examining existing disparities. Forthcoming. *The Extractive Industries and Society, 3*(2), 464–474.

Atuoye, K. N., & Luginaah, I. (2017). Food as a social determinant of mental health among household heads in the Upper West Region of Ghana. *Social Science & Medicine, 180*, 170–180.

Berardi, N., & Marzo, F. (2017). The elasticity of poverty with respect to sectoral growth in Africa. *Review of Income and Wealth, 63*(1), 147–168.

Bisung, E., & Elliott, S. J. (2014). Toward a social capital based framework for understanding the water-health nexus. *Social Science and Medicine, 108*, 194–200.

Bisung, E., Elliott, S. J., Abudho, B., Schuster-Wallace, C. J., & Karanja, D. M. (2015). Dreaming of toilets: Using photovoice to explore knowledge, attitudes and practices around water–health linkages in rural Kenya. *Health & Place, 31*, 208–215.

Bisung, E., Dixon, J., & Luginaah, I. (2018). Development: The past, present and future contributions of health geography. In V. A. Crooks, G. J. Andrews, & J. Pearce (Eds.), *Routledge handbook of health geography*. Routledge.

Braimah, J. A., & Rosenberg, M. W. (2021). "They do not care about us anymore": Understanding the situation of older people in Ghana. *International Journal of Environmental Research. Public Health, 18*, 2337.

Braimah, J. A., & Rosenberg, M. W. (2022). An ecological systems analysis of food access barriers and coping strategies adopted by older adults in Ghana. *The Canadian Geographer/Le Géographe canadien, 66*(1), 107–118.

Braimah, J. A., Sano, Y., Atuoye, K. N., & Luginaah, I. (2019). Access to primary health care among women: The role of Ghana's community-based health planning and services policy. *Primary Health Care Research & Development, 20*(e82), 1–7.

Defo, B. K. (2014). Demographic, epidemiological, and health transitions: are they relevant to population health patterns in Africa? *Global Health Action, 7*(1), 22443.

Dery, F., Achore, M., & Bisung, E. (2022). "Today men's orientation has changed": Gender and household water and sanitation responsibilities in Ghana. In A. Williams & I. Luginaah (Eds.), *Geography, health and sustainability: Gender matters globally*. Routledge.

Dummer, T. J. B. (2008). Health geography: Supporting public health policy and planning. *CMAJ, 178*(9), 1177–1180. https://doi.org/10.1503/cmaj.071783

He, W., Aboderin, I., & Adjaye-Gbewonyo, D. (2020). *Africa aging: 2020* (International Population Reports, P95/20-1). Washington, D.C.

Holmes, M. D., Dalal, S., Volmink, J., Adebamowo, C. A., Njelekela, M., Fawzi, W. W., et al. (2010). Non-communicable diseases in sub-Saharan Africa: The case for cohort studies. *PLoS Medicine, 7*(5), e1000244.

Kahn, K., Tollman, S. M., Collinson, M. A., Clark, S. J., Twine, R., Clark, B. D., et al. (2007). Research into health, population and social transitions in rural South Africa: Data and methods of the agincourt health and demographic surveillance system 1. *Scandinavian Journal of Public Health, 35*(69 suppl), 8–20.

Kangmennaang, J., Bisung, E., & Elliott, S. J. (2020). 'We are drinking diseases': Perception of water insecurity and emotional distress in urban slums in Accra, Ghana. *International Journal of Environmental Research and Public Health, 17*(890), 1–17. https://doi.org/10.3390/ijerph17030890

Kansanga, M. M., Braimah, J. A., Antabe, R., Sano, Y., Kyeremeh, E., & Luginaah, I. (2018). Examining the association between exposure to mass media and health insurance enrolment in Ghana. *International Journal of Health Planning Management, 33*, e531–e540. https://doi.org/10.1002/hpm.2505

Kearns, R. A. (1994). Progress report Medical geography: Difference. *Progress in Human Geography, 19*(2), 251–259. https://doi.org/10.1177/030913259501900206

Kuuire, V., & Dassah, E. (2020). Place and health. In *International encyclopedia of human geography* (Vol. 10, 2nd ed., pp. 125–128). Elsevier. https://doi.org/10.1016/B978-0-08-102295-5.10412-3

Kuuire, V. Z., Tenkorang, E. Y., Amegbor, P. M., & Rosenberg, M. (2021). Understanding unmet health-care need among older Ghanaians: A gendered analysis. *Ageing & Society, 41*(8), 1748–1769.

Luginaah, I. (2009). Health geography in Canada: where are we headed? *The Canadian Geographer/Le Géographe Canadien, 53*(1), 91–99.

Luginaah, I. N., Bezner-Kerr, R., & Dixon, J. (2015). Introduction. In I. N. Luginaah & R. Bezner-Kerr (Eds.), *The geographies of health and development*. Ashgate.

McQuaid, K., Esson, J., Gough, K. V., & Wignall, R. (2021). Navigating old age and the urban terrain: Geographies of ageing from Africa. *Progress in Human Geography, 45*(4), 814–833. https://doi.org/10.1177/0309132520948956

Moon, G. (2020). Health Geography. In *International encyclopedia of human geography* (Vol. 6, 2nd ed., pp. 315–321). Elsevier. https://doi.org/10.1016/b978-0-08-102295-5.10388-9

Odusola, A., Cornia, G. A., Bhorat, H., & Conceição, P. (2017). *Income inequality trends in sub-Saharan Africa: Divergence, determinants and consequences*. United Nations Development Program.

Oleribe, O. O., Momoh, J., Uzochukwu, B. S. C., Mbofana, F., Adebiyi, A., Barbera, T., et al. (2019). Identifying key challenges facing healthcare systems in Africa and potential solutions. *International Journal of General Medicine, 12*, 395–403.

Rishworth, A., Dixon, J., Luginaah, I., Mkandawire, P., & Prince, C. T. (2016). "I was on the way to the hospital but delivered in the bush": Maternal health in Ghana's Upper West Region in the context of a traditional birth attendants' ban. *Social Science & Medicine, 148*, 8–17.

Rosenberg, M. W. (1998). Medical or health geography? Populations, peoples and places. *International Journal of Population Geography, 4*(3), 211.

Rosenberg, M. (2017). Health geography III: Old ideas, new ideas or new determinisms? *Progress in Human Geography, 41*(6), 832–842. https://doi.org/10.1177/0309132516670054

Sano, Y., Antabe, R., Atuoye, K. N., Braimah, J. A., Galaa, S. Z., & Luginaah, I. (2018). Married women's autonomy and post-delivery modern contraceptive use in the Democratic Republic of Congo. *BMC Women's Health, 18*(1), 49.

Sen, A. (1988). The concept of development. In H. Chenery & T. N. Srinivasan (Eds.), *Handbook of development economics*. Elsevier Science Publishers.

Stewart, F. (2011). *Horizontal inequalities as a cause of conflict. A review of CRISE findings*. World Development Report 2011 Background paper. https://openknowledge.worldbank.org/handle/10986/9126

Tuyisenge, G. (2021). Access to maternal health in regions of Rwanda: A qualitative study. In P. T. Makanga (Ed.), *Practicing health geography: The African context*. Springer.

WHO Centre for Health Development (Kobe, Japan). (2004). A glossary of terms for community health care and services for older persons. Kobe, Japan : WHO Centre for Health Development. https://apps.who.int/iris/handle/10665/68896

WHO. (2012). *Health systems in Africa Community: perceptions and perspectives – A multi-country report*. WHO Regional Office for Africa, Brazzaville, Republic of Congo.

Woods, H., Haruna, U., Konkor, I., & Luginaah, I. (2019). The influence of the Community-based Health Planning and Services (CHPS) program on community health sustainability in the Upper West Region of Ghana. *The International Journal of Health Planning and Management, 34*(1), e802–e816. https://doi.org/10.1002/hpm.2694

Part I
Non-communicable Diseases

Chapter 2
A Multilevel Analysis of Neighborhood Inequalities and Non-communicable Disease Multimorbidity in Ghana

Vincent Kuuire, Kilian Atuoye, Elijah Bisung, and Joseph Asumah Braimah

Introduction

The rising burden of non-communicable diseases (NCDs) is a global health crisis (Bennett et al., 2018; Nishtar et al., 2018). In low- and middle-income countries (LMICs), multiple changes to lifestyles, social and demographic composition, environment, climate, and globalization trends are recreating unique 'place' dynamics and exacerbating varied forms of inequalities. These changes—which usually manifest in urban neighborhoods—have implications for multiple disease burdens, including NCD risks (Beaglehole et al., 2011; Frenk & Gomez-Dantes, 2011). Estimates from the World Health Organization Global Burden of Disease Study show NCDs accounted for 72% of deaths in 2016, and this is projected to increase to 77% by 2030 and further up to 81% by 2045 (World Health Organization,

V. Kuuire (✉)
Department of Geography, Geomatics & Environment, University of Toronto Mississauga, Toronto, ON, Canada

Social and Behavioral Health Sciences Division, Dalla Lana School of Public Health, University of Toronto – St. George, Toronto, ON, Canada
e-mail: vincent.kuuire@utoronto.ca

K. Atuoye
Department of Global Development Studies, Queen's University, Kingston, ON, Canada
e-mail: kilian.atuoye@queensu.ca

E. Bisung
School of Kinesiology and Health Studies, Queen's University, Kingston, ON, Canada
e-mail: elijah.bisung@queensu.ca

J. A. Braimah
Department of Health & Society, University of Toronto Scarborough, Toronto, ON, Canada
e-mail: j.braimah@utoronto.ca

J. A. Braimah et al. (eds.), *Health Geography in Sub-Saharan Africa*, Global Perspectives on Health Geography, https://doi.org/10.1007/978-3-031-37565-1_2

2018a, b). These estimates also revealed disparities across geographical regions and wealth categories. For instance, while high-income countries are projected to record a 15% decline in NCD mortality rate by 2030 from 346.2 per 100,000 in 2016, low-income countries will experience a further rise from 631.5 to 633.8 per 100,000 over the same period, more than doubling the gap in NCD mortality between the two groups of countries (World Health Organization, 2018b).

These regional disparities in NCD burden underscore the long-held position that health is a function of 'place'—a space of varying degrees of livability and inequalities, including neighborhoods (Cummins et al., 2007; Curtis, 2016; Kearns, 1993; Rosenberg, 1998). In sub-Saharan Africa (SSA), lingering infectious diseases (e.g., HIV, tuberculosis, and malaria), maternal, neonatal, and nutritional diseases (i.e., diseases of poverty) are co-occurring with NCDs, leading scholars to describe this multiple vulnerability as a 'double burden of disease' (Agyei-Mensah & de-Graft Aikins, 2010; Bygbjerg, 2012). Others note a 'triple burden of diseases' by drawing attention to the impact of emerging diseases from the effects of globalization and climate change (Frenk & Gomez-Dantes, 2011). By and large, the epidemiological transition theory, which suggests a shift in disease profile from infectious to non-infectious diseases as societies experience demographic and socioeconomic changes, seems not to hold in SSA (Agyei-Mensah & de-Graft Aikins, 2010; Santosa et al., 2014). These complicated disease co-occurrences not only show the enormity of disease burden in the region but also highlight the challenge with the binary conceptualization of communicable versus non-communicable diseases or isolating diseases into single categories in empirical analysis and policy.

Nonetheless, global actions on NCD burden have centered on broader society-level dynamics with a strong emphasis on shared risk factors of the most prevalent NCDs—tobacco and alcohol use, unhealthy diet, and physical inactivity (Nishtar et al., 2018; Shrivastava et al., 2016). For example, the World Health Organization's Global Action Plan for the Prevention and Control of NCDs, as well as the United Nations Political Declaration on the Prevention and Control of NCDs, focus on the implementation of punitive tax policies on alcohol and tobacco use, salt consumption regulation, promotion of physical activity, healthy eating, as well as early detection and treatment of NCDs (Beaglehole et al., 2011; World Health Organization, 2013). The Gaborone call to action on NCDs in Africa also emphasizes lifestyle approaches, highlighting the need to modify social, cultural, and behavioral risk factors by investing in research as part of a six-point agenda (Munodawafa, 2019). Given these narrow approaches in addressing co-occurrence and multiple burdens of NCDs, the Global Alliance for Chronic Diseases recommends a shift in focus to the application of the multimorbidity approach in NCDs research and policy (Hurst et al., 2018). The concept of NCDs multimorbidity allows for the analysis of two or more NCD co-occurrences, recognizing the multiple etiology, occurrences, burden, and risks of NCDs (Hurst et al., 2018; Morgan et al., 2019).

Obviously, the characterization of NCD risks as stemming from physiological and behavioral factors over the years has glossed over the importance of 'place' (Adjaye-Gbewonyo & Vaughan, 2019). The concept of 'place' embraces the physical, socioeconomic, cultural, political, and environmental characteristics and their interaction with individuals and populations in a specific locality (Cummins et al., 2007; Curtis, 2016; Kearns, 1993; Rosenberg, 1998). These interactions, along with the influence of outside factors, including globalization and climate change, contribute to changing dynamics of 'place'—its identity and character. Thus, Curtis (2016) explained that 'place' as an embodiment of multiple factors can directly influence the health of individuals, as well as modify how other context-specific factors (e.g., socioeconomic inequalities) affect health. For this reason, NCD risks can be decomposed into three interacting groups of variables: place identity (e.g., perception of inherent place vulnerability and socio-cultural image); place-based factors (e.g., access to healthcare, water security, poverty gap, and income inequality); and individual level factors (e.g., physiological, behavioral, and socioeconomic characteristics).

However, studies that examined 'place' in relation to NCD burden in SSA have mainly focused on describing the clustering of specific NCDs. For example, Stanifer and colleagues examined clustering of chronic kidney disease, hypertension, glucose impairment, and obesity in urban and rural communities in northern Tanzania (Stanifer et al., 2016). While their study showed different clustering levels for hypertension and glucose impairment across neighborhoods, it failed to examine how 'place' interacts with other factors to shape disparities in the clustering of these diseases. Similarly, in their study of NCDs mortality and risk factors in formal and informal neighborhoods in Ouagadougou, Rossier and colleagues did not capture the interactive dynamics of place level effects (Rossier et al., 2014). As the concept of NCD multimorbidity and place effects have received limited attention, there is low understanding of how 'place' vulnerabilities directly pose risk to NCDs, as well as mediate the effects of other social determinants of NCDs. In contributing to the limited theoretical and empirical attention in this area, we take the position that analysis of NCD risk factors should employ an integrative approach, embracing both the direct and modifying effect of 'place' on NCDs. Advancing this approach is critical to revealing the complexities of NCD burden and provides crucial empirical evidence to support a more encompassing NCD policymaking that addresses both individual level factors and place inequalities.

Contextualizing Place and Social Determinants of Health (SDH)

The World Health Organization's Global Action Plan for the Prevention and Control of NCDs emphasizes a life course approach, describing it as an inclusive theoretical framework for analyzing and addressing NCDs among all ages (World Health Organization, 2013). The life course approach assumes that an individual's health status is influenced by a combination of lifelong factors going back to childhood and even before birth (Lynch & Smith, 2005; Mikkelsen et al., 2019). In effect, a life course approach takes into consideration the historical context of an individual, which plays out over complex interactions of cultural, socioeconomic, and political factors within neighborhoods.

However, to fully integrate place dynamics into analysis of NCDs, Marmot and Bell suggest a shift to social determinants of health (SDH) with attention on life-long experiences (Marmot & Bell, 2019). Sanctioned by the World Health Organization Commission, the SDHs framework emphasizes the crucial role of social determinants in the patterns and disparities of NCDs associated disability and mortality (Marmot & Bell, 2019). Indeed, Marmot and Bell describe social determinants as consisting of the "'causes of the causes' of health inequality" (Marmot & Bell, 2019, p. 10). By this, disparities in risks to NCDs relate to the unequal structural and neighborhood conditions in which people are "born, grow, live, work and age," which in themselves are shaped by the distributional effect of power, money, and other resources (Marmot et al., 2008, p. 1161). In addition to place dynamics, SDH recognizes that individuals with pre-existing health conditions and other biological predispositions have higher risks of NCDs (Kawachi et al., 1999; Marmot & Bell, 2019). Given the utility of SDH in explaining health inequalities, it has been applied in several studies across disciplinary and geographical boundaries. For instance, SDH is employed as a framework of analysis to highlight urban slums' vulnerability to NCDs in India (Lumagbas et al., 2018). In the specific context of SSA, SDH has been applied in studying neighborhood effect on NCDs in Tanzania (Stanifer et al., 2016) and in contextualizing progress and challenges in addressing chronic NCDs in South Africa (Puoane et al., 2012). In this study, we applied SDH in analyzing how 'place' dynamics combine with individual level factors to influence the risk of NCD multimorbidity in Ghana.

Methods

Study Context

The study was conducted in six local government administrative areas drawn from the Greater Accra and Northern regions in Ghana. The country's population in 2019 was estimated at 30 million, with women forming a slight majority (50.8%). Individuals aged 18 years and above form 55%, and with a median age of 21, the country has a very youthful population (Ghana Statistical Service, 2020). The urban population has been fast-growing, jumping from 23% in 1960 to 51% in 2010, while the proportion of individuals employed in the services sector, often associated with sedentary lifestyles, formed 46% of the population (Ghana Statistical Service, 2012, 2014). Since the economic meltdown of the 1980s, Ghana has recorded progress in economic growth, allowing for investment in critical social services, including education and health. In particular, the implementation of a free compulsory universal basic education policy in the early 1990s led to the progressive expansion of education access for most children (Akyeampong, 2009), while the national health insurance policy which started in 2004 also contributed to a significant improvement in health care access (Antabe et al., 2019; Aryeetey et al.,

2016). The impacts of these policies are observed in many ways, including an improvement in life expectancy at birth, from 52.3 years in 1980 to 63.8 years in 2018 (World Bank, 2020). In addition, poverty incidence has plummeted in the last three decades. The proportion of people living below the international poverty line (living on $1.90 a day) declined from 52.6% to 21.4% between 1991 and 2012, and further down to 11.9% in 2017 (Ghana Statistical Service, 2018).

Despite these modest achievements, poverty and income inequalities have been rising. As highlighted by the 2019 Human Development Report, growth in incomes of the bottom 40% was 24% less than the national average between 1995 and 2015, which suggest a widening disparity as the country's economy expands into the middle-income bracket (UNDP, 2019). In addition, with less than the corresponding rise in infrastructural, employment opportunities and population increase, urban localities have become hotspots for growing inequalities. Access to portable water reduced by 22.5% in the last decade alone (2000–2010), agricultural employment for the majority of unskilled individuals shrunk by 21% between 1992 and 2010, while the housing deficit is estimated to reach 2 million in 2020 (World Bank, 2015). Given these growing inequalities and vulnerabilities alongside increasing risk from fast changing environmental, demographic, and socioeconomic dynamics in the country, it is not surprising that NCD burden is turning out to be a major health crisis. In 2016, NCDs accounted for 36% of disability adjusted life years (DALYs)—years lost to early death and morbidity, an increase by 14 percentage points since the year 2000, while communicable, maternal, perinatal, and nutritional conditions declined by 17 percentage points from 71% over the same period (see Fig. 2.1).

Among NCDs, cardiovascular diseases (CVDs) accounted for most mortality and morbidity cases in 2016 (9% of all DALYs), followed by mental and substance use abuse (4%), cancers (3%), diabetes (2%), and chronic respiratory diseases (2%). The data also shows the rising importance of other NCDs, including digestive, sense

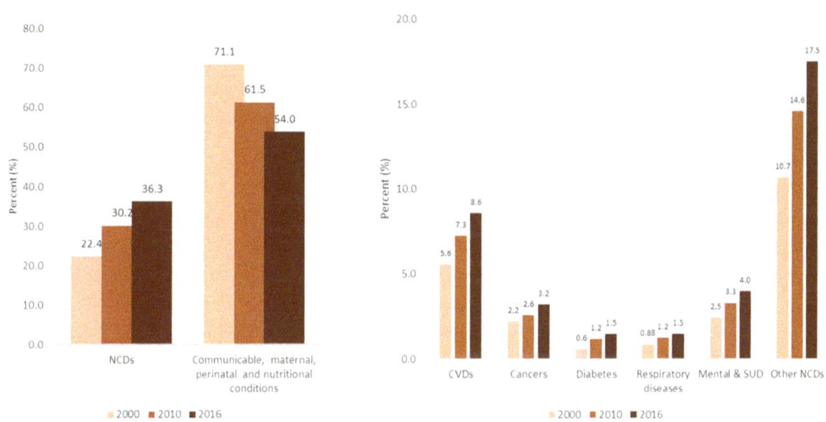

Fig. 2.1 Causes of disability adjusted life years in Ghana, derived from WHO Global Health Estimates (2000–2016)

organ, congenital, and Genitourinary diseases (World Health Organization, 2018a). Amidst these worrying trends, scholars and policymakers alike have bemoaned the haphazard policy orientation for NCDs, low investment in health promotion and infrastructure/personnel for NCD case management, and paucity of data on NCDs (Laar et al., 2019; Ministry of Health, 2018; Tenkorang & Kuuire, 2016).

Data

Data for this study came from a cross-sectional study conducted among individuals aged 18 and above (n = 1145) in six local government areas in Ghana—Aboabu/Sabonjida, Gumani, and Vitting in the Northern region: and Korle-Gonno, Madina, and Teshie in the Greater Accra region. We randomly selected individuals from a predefined sample size relative to the population of study localities. While this approach ensured data randomness, there was clustering around the study sites. Data collection used standardized instruments that covered self-reported cases of non-communicable and infectious diseases; self-assessed health status; lifestyle patterns (i.e., physical activity, eating habit, level of alcohol consumption, and tobacco use); demographic; and socioeconomic characteristics of individuals. We included neighborhood-level inequality variables such as the Gini index, water security, poverty gap, and access to health care from the Ghana Statistical Service (GSS). Data were collected between June and August 2019. Twelve experienced enumerators were recruited and trained on the data collection instruments and ethical considerations guiding the study. These enumerators were deployed in pairs to each of the study sites. The lead researcher personally supervised the data collection. Data errors were significantly reduced largely because of the quality of enumerators and supervision during data collection. The data was processed in IBM SPSS version 21. Ethical approval for the study came from the Non-Medical Research Board of the University of Toronto.

Measures

NCDs multimorbidity is the outcome variable in this study, a three categorical variable derived from fourteen 'yes/no' response questions on reported NCDs[1] (coded 1 = no NCD, 2 = single NCD, and 3 = multiple NCDs). Neighborhoods from which data was drawn formed the cluster level variable.

[1] Reported NCDs were arthritis, diabetes, hypertension, stroke, angina, chronic lung disease, asthma, cataract, oral health problem, injuries, cervical cancer, breast cancer, prostrate cancer, and any other chronic disease.

Informed by the social determinants of health and literature on NCD risk factors (Marmot & Bell, 2019), we included three blocs of variables as covariates—self-rated health (SRH) and lifestyle factors; socioeconomic and demographic factors; and neighborhood inequalities. SRH was constructed from two 5-likert questions on physical and mental health, dummied into poor and good health. Lifestyle variables included tobacco use (1 = never smoked, 2 = ever smoked, 3 = currently smoking), physical activity (1 = active, 2 = fairly active, 3 = not active), dietary choice outside the home (1 = local only, 2 = local and foreign, 3 = foreign only), and consumption of unhealthy diet/drink commonly classified as 'junk foods and drinks' (1 = yes, 2 = no). Socioeconomic and demographic factors were age, gender, marital status, educational attainment, employment status, and household wealth quintile. Wealth quintiles were constructed from household assets and building characteristics using Principal Component Analysis with an acceptable scale reliability (α = 0.78). Neighborhood factors were perceived healthcare status derived from perceptions about availability, accessibility, and quality of healthcare services; Gini index; water security; and poverty gap obtained from GSS.

Analytical Technique

We employed descriptive statistics and multivariate regression to examine how place-based and individual-level factors influence the burden/risk of NCD multimorbidity in Ghana. Descriptive statistics showed sample distribution, and with inferential statistics (Pearson chi2 and Cramer's V) we identified significant bivariate associations between our predictor variables and NCD multimorbidity. In the multivariate analysis, we fitted mixed-effects models using multilevel multinomial logistic regressions to accommodate the three categorical non-ordered nature of the outcome variable—NCD multimorbidity—and account for the multistage correlation structure in our data. The models estimated a fixed effect component representing how individual level and place-based factors independently contribute to NCD multimorbidity risk, and a random effect component accounting for the proportion of variance in NCD multimorbidity burden explained by place. We first fitted a random intercept only model (empty model) to examine whether our data justified the analysis of random effect at the neighborhood level. We then built four additional models, with the first three adjusting for one of the three blocs of covariates while the last accounted for all covariates. Specifically, we accounted for neighborhood inequalities in model 2, predisposing health status and lifestyle factors in model 3, socioeconomic and demographic factors in model 4, and all covariates in model 5.

Regression coefficients were transformed into relative risk ratios (RRR) for easy and more informative interpretation. RRR greater than 1 is interpreted as higher risk of reporting multiple NCD or single NCD than reporting no NCD, and RRR less than 1 is interpreted as lower risk of reporting multiple NCD or single NCD than no NCD. Variance partition coefficients assessing the proportion of variance of NCD multimorbidity explained by place morphology/identity was computed using the formulae:

$$\text{VPCn} = \frac{\sigma_n^2}{\sigma_n^2 + \pi^2 / 3} \tag{2.1}$$

where σ_n^2 represents cluster level variance (i.e., neighborhood of residence); and $\pi^2 / 3$ represents variance within individuals (Goldstein et al., 2002).

Results

Sample Characteristics

As shown in Fig. 2.2, individuals in the study were fairly distributed across the six neighborhoods. While about half (54%) of the individuals in the study were resident in Aboabu/Sabonjida, Gumani, and Korle-Gonno (18% in each), 34% were in Target/Vitting and Teshie (17% in each), and 12% were resident in Madina (see Fig. 2.2).

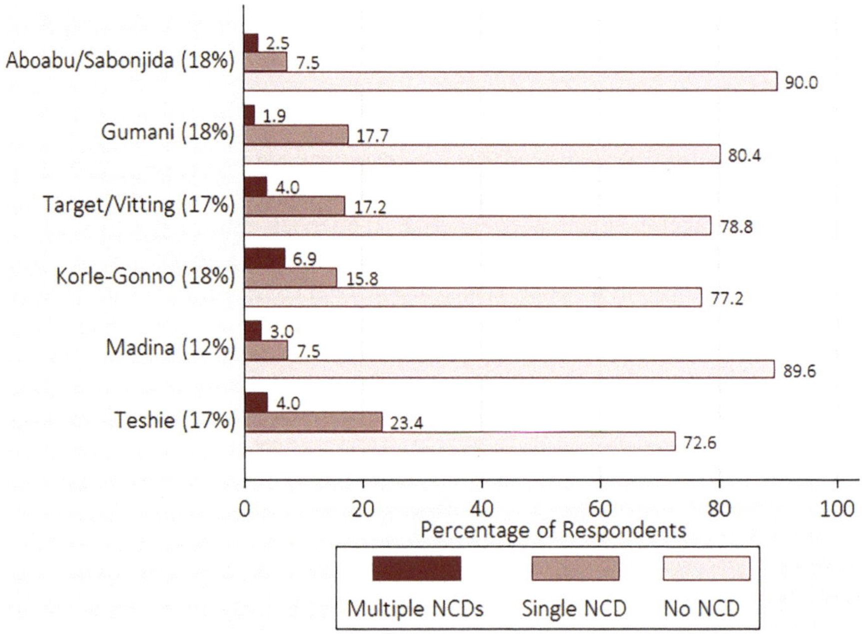

Fig. 2.2 Non-communicable disease multimorbidity by neighborhoods in Ghana (N = 1145)

Neighborhood- and individual-level characteristics are presented in Table 2.1. Approximately 30% of individuals reported living in neighborhoods with poor healthcare conditions, and 72% with poor access to potable water. The average neighborhood income inequality was 35.3, while poverty gap stood at 4.1.

Additionally, a little above half of the individuals (62%) reported good SRH. Also, lifestyle factors reflected the general socio-cultural landscape of the Ghanaian society. For instance, as smoking is socially stigmatized in urban Ghana, we found that 91% of individuals had never smoked, and approximately 4% who ever used tobacco had reported abandoning it at the time of the study. Similarly, while 40% were physically inactive, 57% had the habit of choosing non-traditional diets when eating outside, and approximately 83% reported frequent consumption of food commonly classified as 'junk foods and drinks.' We also observed vast disparities in socioeconomic and demographic characteristics. Most individuals were youthful, with only 13% of them above 54 years. Additionally, more than half were males (56%), married (55%), attained basic education (54%), and had temporary employment (66%). Wealth was evenly distributed in quintiles with individuals in the middle wealth category representing 20%. In this study, 15% of individuals reported single NCD, and 4% reported multiple NCD.

Descriptive Statistics

Descriptive statistics showing the bivariate association between predictors and NCD multimorbidity are presented in Table 2.1 and Fig. 2.2. The study found wide disparities in the burden of NCDs across the six neighborhoods (see Fig. 2.2). For example, while the highest burden of NCD multimorbidity was reported in Korle-Gonno (7%), the lowest was in Gumani (2%). Similarly, Teshie (23%) had the highest proportion of single NCD burden, while the lowest was reported in Madina (8%) and Aboabu/Sabonjida (8%). Overall, Aboabu/Sabonjida had the lowest burden of NCDs.

Given these disparities, it is not surprising that neighborhood-level characteristics such as availability of healthcare and income inequality were statistically associated with NCD multimorbidity (see Table 2.1). A greater proportion of individuals who reported multiple NCDs (5%) and single NCD (20%) were residents in neighborhoods with poor healthcare services. Surprisingly, the study found an inverse relationship between income inequality and the burden of NCD multimorbidity. Low-income inequality neighborhoods (mean Gini coefficient = 34.8) were associated with high prevalence of multiple NCDs, while neighborhoods with high-income inequality (mean Gini coefficient = 35.4) rather reported a low burden of NCD multimorbidity.

Additionally, the risk of predisposing health conditions and lifestyle factors for NCD multimorbidity were confirmed in this study. For instance, whereas 6% of individuals with inactive lifestyles reported multiple NCDs and 20% reported single NCD, individuals with active lifestyles had a relatively lower burden—4% for multiple NCDs and 7% for single NCD. Similarly, individuals with the habit of

Table 2.1 Distribution of NCD multimorbidity by selected study variables (n = 1145)

Variables	Total sample	NCD multimorbidity: Multiple NCDs	Single NCD	No NCD	Inferential statistics
	%/ x̄ (std)	%/ x̄ (std)	%/ x̄ (std)	%/ x̄ (std)	
Perceived healthcare status					
Poor	31.35	4.46	20.06	75.49	χ^2 = 913.8; Pr = 0.008 Cramer's V = 0.08
Fair	8.21	5.32	18.09	76.60	
Good	60.44	3.18	12.43	84.39	
Water security					
No	71.70	3.41	14.37	82.22	χ^2 = 3.06; Pr = 0.216 Cramer's V = 0.05
Yes	28.30	4.63	17.59	77.78	
Gini index[a]	35.33(1.81)	x̄ =34.8(1.89)	x̄ =35.3(1.96)	x̄ =35.4(1.78)	χ2 = 40.4; Pr = 0.000
Poverty gap[b]	4.09(3.34)	x̄ =3.3(3.52)	x̄ =4.0(3.51)	x̄ =4.1(4.30)	χ2 = 43.9; Pr = 0.000
Self-rated health					
Poor	38.08	6.88	20.64	72.48	χ^2 = 37.8; Pr = 0.000 Cramer's V = 0.18
Good	61.92	1.83	11.99	86.18	
Tobacco use					
Never smoked	90.66	3.76	14.93	81.31	χ^2 = 6.6; Pr = 0.157 Cramer's V = 0.05
Stopped smoking	3.76	2.33	9.30	88.37	
Currently smoking	5.59	4.69	25.00	70.31	
Physical activity					
Active	14.41	3.64	6.67	89.70	χ^2 = 25.6; Pr = 0.000 Cramer's V = 0.11
Fairly active	45.94	2.28	14.07	83.65	
Not active	39.65	5.51	19.82	74.67	
Dietary choice outside home					
Local food only	35.28	2.23	11.14	86.63	χ^2 = 17.7; Pr = 0.001 Cramer's V = 0.09
Local and foreign	7.51	3.49	10.47	86.05	
Foreign only	57.21	4.73	18.47	76.79	
Consumption of unhealthy foods/drinks					
Yes	83.49	3.35	15.48	81.17	χ^2 = 2.71; Pr = 0.253 Cramer's V = 0.05
No	16.51	5.82	14.29	79.89	

(continued)

Table 2.1 (continued)

Variables	Total sample %/ x̄ (std)	NCD multimorbidity Multiple NCDs %/ x̄ (std)	Single NCD %/ x̄ (std)	No NCD %/ x̄ (std)	Inferential statistics
Age					
55+ years	13.10	12.67	33.33	54.00	$\chi^2 = 110.5$; Pr = 0.000 Cramer's V = 0.22
45–54 years	17.38	5.03	18.09	76.88	
35–44 years	25.59	0.68	15.70	83.62	
25–34 years	31.62	1.66	8.56	89.78	
18–24 years	12.31	4.26	8.51	87.23	
Gender					
Male	55.81	3.60	12.52	83.88	$\chi^2 = 8.8$; Pr = 0.012 Cramer's V = 0.09
Female	44.19	3.95	18.77	77.27	
Marital status					
Formerly married	12.84	8.84	25.85	65.31	$\chi^2 = 45.7$; Pr = 0.000 Cramer's V = 0.14
Currently married	54.93	3.34	17.17	79.49	
Never married	32.23	2.44	7.86	89.70	
Education					
No formal education	20.00	4.80	18.34	76.86	$\chi^2 = 3.7$; Pr = 0.444 Cramer's V = 0.04
Basic education	53.71	3.25	14.15	82.60	
Tertiary education	26.29	3.99	15.28	80.73	
Employment					
Unemployed	16.86	3.63	19.17	77.20	$\chi^2 = 4.9$; Pr = 0.301 Cramer's V = 0.05
Employed	21.57	2.43	13.36	84.21	
Temporarily employed	61.57	4.26	14.89	80.85	
Wealth quintile					
Rich	40.00	5.02	18.12	76.86	$\chi^2 = 9.2$; Pr = 0.055 Cramer's V = 0.06
Middle	20.00	3.06	12.23	84.72	
Poor	40.00	2.84	13.97	83.19	

Note: x̄ =mean; std. = standard deviation
[a]min = 33.0, max = 37.5
[b]min = 0.3, max = 9

opting for foreign meals when eating out reported a higher burden of multiple NCDs (5%) and single NCD (18%) than those who preferred either mixing local and foreign dishes or eating only local meals. While tobacco use and consumption of junk food and drinks showed varying risks for NCD multimorbidity, these disparities were not statistically significant and, therefore, not statistically different from expected at the bivariate level.

Furthermore, demographic characteristics such as age, gender, and marital status were statistically associated with disparities in NCD multimorbidity. Older individuals, females, and individuals who were either divorced or separated had higher than expected burden of both multiple NCDs and single NCD. However, the socioeconomic status of individuals, including educational attainment, employment status, and household wealth, did not matter in NCD multimorbidity disparities from our bivariate analysis.

Multivariate Results

Table 2.2 presents the results from our multilevel analysis of predictors of NCD multimorbidity in Ghana. The main objective of the study was to examine the proportion of variance in NCD multimorbidity risk attributable to 'place'—the physical, environmental, political, socioeconomic, and cultural embodiment of neighborhoods, and how individual-level and place-based factors contribute to NCD multimorbidity disparities. The analysis in our null model revealed that approximately 5.7% of the variance in NCD multimorbidity is explained by 'place.' After introducing neighborhood factors in model 2, we found that perceived healthcare status was a significant predictor of NCD multimorbidity, but the statistical association between NCD multimorbidity and income inequality (Gini coefficient) and poverty gap observed in the bivariate analysis disappeared. In contrast with neighborhoods with poor healthcare conditions, those with good healthcare conditions had lower risk of single NCD (RRR = 0.38, $p \leq 0.05$) and multiple NCDs (RRR = 0.40, $p \leq 0.01$). Similarly, neighborhoods with healthcare conditions rated as fair had their risk for single NCD reduced by 45% but shared the same risk level for multiple NCDs risk with neighborhoods with poor health conditions. By accounting for neighborhood characteristics, the proportion of variance in NCD multimorbidity explained by 'place' increased to 6.6%.

The association between SRH/lifestyle factors and NCD multimorbidity remained robust in model 3. We found that individuals with good SRH had lower risk of single and multiple NCDs (RRR = 0.52, $p \leq 0.05$; and RRR = 0.24, $p \leq 0.05$, respectively). Also, sedentary lifestyle (physically inactive) was associated with higher risk of single NCD (RRR = 2.04, $p \leq 0.05$), while preference for foreign meals when eating outside had higher risk for both single and multiple NCDs (RRR = 1.95, $p \leq 0.01$; and RRR = 2.57, $p \leq 0.01$, respectively). The proportion of NCD multimorbidity variance explained by 'place' in model 3 was 6.3%.

Moreover, we found after accounting for socioeconomic and demographic factors together in model 4 that while age, gender, and household wealth maintain their significance from bivariate analysis, marital status was no longer statistically associated with NCD multimorbidity. In particular, younger individuals showed lower risk of reporting single and multiple NCDs compared with individuals aged

Table 2.2 Multivariate multilevel multinomial logistic regression estimates of non-communicable diseases (NCDs) risk factors in Ghana (n = 1145)

		Neighborhood		Lifestyle		Socioeconomic and demographic		Full model	
		Single	Multiple	Single	Multiple	Single	Multiple	Single	Multiple
Variables	Intercept	RRR (RSE)	RRR (RSE)	RRR (RSE)	RRR (RSE)	RRR (RSE)	RRR (RSE)	RRR (RSE)	RRR (RSE)
Perceived healthcare status (ref: poor)									
Fair		0.55(0.15)*	0.73(0.46)					0.58(0.19)	0.89(0.69)
Good		0.38(0.14)*	0.40(0.14)**					0.43(0.16)*	0.39(0.19)*
Water security (ref: no)									
Yes		1.09(0.35)	1.29(0.27)					1.65(0.61)	2.31(1.00)
Gini index		0.96(0.31)	0.83(0.13)					0.92(0.23)	0.72(0.04)***
Poverty gap		0.95(0.18)	0.96(0.10)					1.04(0.17)	1.06(0.06)
SRH (ref: poor)									
Good				0.52(0.17)*	0.24(0.13)*			0.55(0.10)***	0.26(0.09)***
Tobacco smoking (ref: never smoked)									
Stopped smoking				1.02(0.67)	0.85(1.02)			0.80(0.69)	0.43(0.73)
Currently smoking				1.99(0.70)	1.07(0.35)			1.24(0.58)	0.43(0.25)
Physical activity (ref: active)									
Fairly active				1.73(0.75)	0.51(0.33)			1.46(0.63)	0.50(0.21)*
Not active				2.04(0.73)*	0.95(0.79)			1.27(0.47)	0.56(0.40)
Dietary choice outside home (ref: local food only)									
Local + foreign				1.16(0.36)	2.80(2.72)			1.11(0.32)	3.24(2.77)
Foreign only				1.95(0.48)**	2.57(0.85)**			1.42(0.24)*	1.82(0.56)*
Consumption of unhealthy foods/drinks (ref: yes)									
No		0.70(0.33)	1.34(0.20)					0.56(0.20)*	1.35(0.31)
Age (ref: 55+)									
45–54						0.38(0.13)**	0.31(0.14)**	0.43(0.22)	0.40(0.22)

(continued)

Table 2.2 (continued)

Variables	Intercept	Neighborhood		Lifestyle		Socioeconomic and demographic		Full model	
		Single	Multiple	Single	Multiple	Single	Multiple	Single	Multiple
		RRR (RSE)	RRR (RSE)	RRR (RSE)	RRR (RSE)	RRR (RSE)	RRR (RSE)	RRR (RSE)	RRR (RSE)
35–44						0.29(0.06)***	0.04(0.02)***	0.33(0.12)**	0.06(0.02)**
25–34						0.16(0.09)***	0.10(0.07)**	0.16(0.11)**	0.15(0.13)*
18–24						0.18(0.11)**	0.28(0.29)	0.20(0.14)*	0.66(0.68)
Gender (ref: male)									
Female						1.76(0.34)**	1.34(0.46)	1.68(0.36)*	1.01(0.33)
Marital status (ref: formerly married)									
Currently married						0.94(0.23)	0.55(0.24)	0.93(0.26)	0.66(0.23)
Never married						0.73(0.38)	0.51(0.28)	0.82(0.43)	0.58(0.46)
Education (ref: no formal education)									
Basic education						1.13(0.29)	0.97(0.43)	1.18(0.28)	0.65(0.24)
Tertiary education						1.30(0.35)	1.23(0.88)	1.21(0.35)	0.86(0.72)
Employment (ref: unemployed)									
Employed						0.62(0.22)	0.72(0.42)	0.71(0.27)	0.85(0.50)
Temp. employed						0.79(0.18)	1.24(0.19)	0.85(0.25)	1.26(0.20)
Wealth quintile (ref: rich)									
Middle						0.62(0.15)	0.49(0.09)**	0.63(0.15)	0.40(0.06)***
Poor						0.56(0.18)	0.43(0.24)	0.55(0.20)	0.34(0.20)
Random effects									
Variance at neighborhood level	0.20	0.22		0.18		0.23		0.18	
VPC at neighborhood level	0.057	0.063		0.052		0.066		0.053	

*Note***:** multinomial logit reference category = no NCD; *RRR* relative risk ratio, *RSE* robust standard errors, *VPC* variance partition coefficient
$*p \leq 0.05$, $**p \leq 0.01$, $***p \leq 0.001$

55 and above. Also, despite sharing the same risk level with males for multiple NCDs, females had higher risk for single NCD (RRR = 1.76, $p \leq 0.01$). In contrast, the risk for multiple NCDs for individuals from households in the middle wealth category (average wealth) was approximately half of that of their colleagues from rich households (RRR = 0.49, $p \leq 0.01$). The effect of socioeconomic and demographic factors on the influence of 'place' on NCD multimorbidity was 6.6%.

Accounting for all covariates in model 5 revealed other important nuances in the association between place-based factors and NCD multimorbidity. Most importantly, income inequality (Gini index) which was not significant in model 2, appeared significant and had an inverse relationship with risk of NCD multimorbidity. Meanwhile, access to good healthcare is further confirmed in model 5 to lower the risk of NCD multimorbidity. Similarly, the significance of SRH/lifestyle factors were robust; only physical activity showed a slight disparity. Being fairly active (moderate physical activity), which appeared not significant in model 3, was now associated with lower risk of multiple NCDs (RRR = 0.50, $p \leq 0.05$). Also, age, gender, and wealth maintained a significant relationship with NCD multimorbidity. Overall, the results showed that 18% of the variance in NCD multimorbidity was explained by the neighborhood of residence. This represents a decline of 2% with respect to the null model. 'Place' explained 5.3% of NCD multimorbidity variance in model 5. Changes in variance partition coefficient (VPC) values across models indicate the varying influence of the three blocs of covariates, whereby smaller VPC values suggest the bloc of variables in the model had a greater influence than others. Thus, VPC of 0.052 in model 3 implies that SRH and lifestyle factors slightly have a greater influence on NCD multimorbidity than socioeconomic, demographic, and place-based factors.

Discussion

In this study, we examined NCD risk factors in Ghana and the influence of neighborhood inequalities. The rising global burden of NCDs has been a public health concern with a disproportionately unbearable impact on the fragile and scarce economic and health resources in low-income countries. The Sustainable Development Goals (SDGs) have recognized how a reduction of mortality from NCDs even by one-third is a critical step toward ensuring healthy lives and promoting well-being for all (Bennett et al., 2018; Munodawafa, 2019; Nugent et al., 2018). Achieving this global target is a collaborative process that integrates other related targets on poverty, food and nutritional security, maternal and child health, infectious diseases, gender, and climate change (Nishtar et al., 2018; Nugent et al., 2018). As these multiple factors converge in local contexts (neighborhoods), addressing NCD burden requires context-specific policies and programs that take into consideration the uniqueness and complexity of NCDs and their risk factors. This study provides important insight into place-based and individual-level NCD multimorbidity risk factors for the attention of policy in Ghana and similar contexts in SSA.

The study confirms that place matters in NCD risk. We found 'place' effect as being more complex than often conceptualized in NCD risk in contexts such as SSA. Prior studies have applied clustering as a proxy for the influence of 'place' (Haregu et al., 2018; Rossier et al., 2014; Stanifer et al., 2016). For example, Rossier et al.'s analysis of NCDs clustering in Burkina Faso underscored how formal neighborhoods have become zones for overweight, and therefore, pose a higher risk to NCDs (Rossier et al., 2014). Others have used urbanicity and urban-rural divide as an estimation technique to examine place effect on NCDs, often revealing urban settings as high-risk localities (Osman et al., 2019; Pelzom et al., 2017; Riha et al., 2014). Neighborhood effect, which focuses on locational attributes—civil infrastructure, walkability function, and greenness—appears as another research domain highlighting 'place' effect on NCDs (Adhikari et al., 2019; Iyer et al., 2021). Common across these research domains is the assumption that place-based characteristics adequately represent the complexities of place and its effect on NCDs. Our analyses demonstrate that place is multidimensional, which for the sake of analysis can be peeled off into three intricately nested layers—individuals, place-based, and factors contributing to place identity (e.g., historical, political, contemporary global dynamics). While scholarship tying health to place emphasizes complex interactions of place-based and individual-level factors (Cummins et al., 2007; Curtis, 2016; Kearns, 1993; Rosenberg, 1998), we found the third layer of place alone accounting for more than 5% of the variance in NCD risk in this study. Although relatively small, the finding advances our understanding of how 'place' contributes to the burden of NCDs beyond place-based factors. Importantly, it makes a case for the study of place identity as an important factor in NCD burden in SSA, especially addressing questions on elements of place identity that matter on NCD burden, how these factors can be measured and quantified in relation to the complex interaction of factors in local neighborhoods.

Furthermore, the finding that neighborhoods with high-income inequality (Gini index) are associated with lower risk of NCDs may stand contrary to the dominant literature, but it is not entirely surprising. Scholars have suggested a mediation effect between epidemiological transition and gross national income which influences how socioeconomic inequalities affect disparities in NCD burden (Ezzati et al., 2005; Osman et al., 2019; Spiteri & von Brockdorff, 2019). For instance, Spiteri and von Brockdorff concluded from their study of eleven countries that socioeconomic inequalities hold a 'U'-shaped relationship with NCD risks (Spiteri & von Brockdorff, 2019). They posited a positive relationship between high-income inequalities and low risk of NCDs in countries with GDP below a certain threshold, and beyond which the relationship reverses. This is possible because structures (e.g., health and employment) in developing countries are transitory, in the process of evolving into a more formalized system. Thus, unlike high-income countries with well-developed uni-structural systems, developing countries have bi-structural societies, generally representing a socioeconomic divide. Both structured societies have distinct pools of resources, employment types, and consumption patterns. In Ghana and most parts of SSA, wealthy individuals are active participants/members of the emerging economic structure that is characterized by over consumption and sedentary lifestyles. In contrast, poor individuals generally engage in more active working lifestyles in agriculture, craftwork, and petty trade/businesses. These individuals are also without the luxury of accessing non-traditional meals, often in the domain of fast and junk foods, but at the same time enjoy strong social capital known to reduce NCD risk (Hu et al., 2014; Moore & Kawachi, 2017). Therefore, neighborhoods with high income inequalities represent a wide divide between the few wealthy individuals with higher risk and the vast majority of individuals with lower risk of multiple NCDs. It is, however, important to note from our findings that disparities in risk of single NCD are less apparent in the income inequality divide, suggesting that structural income inequality effect in neighborhoods is more pronounced in multiple NCD burden.

Moreover, good neighborhood healthcare condition is associated with lower risk for both single and multiple NCDs. This finding is consistent with the evidence that access to good and quality healthcare promotes healthy lives and reduces the burden of NCDs (Nishtar et al., 2018; Nugent et al., 2018). Investing in healthcare preparedness for NCD prevention, early diagnosis, and treatment is a proven and enduring action on NCDs. Indeed, Nugent et al. have positioned such investment as the bridge to the attainment of ten other SDGs (Nugent et al., 2018). In Ghana, the introduction of a national health insurance policy in the last fifteen years has drastically reduced wealth disparities in access to primary healthcare (Antabe et al., 2019; Aryeetey et al., 2016). Together with vigorous implementation of community-based health planning and service programs, basic healthcare is brought to the doorsteps of people in several neighborhoods. Anfaara et al. observed a positive impact of these health policies on healthy behaviors against hepatitis, which directly relate to positive health behaviors and general consciousness for healthy living (Anfaara et al., 2018). It is also possible that Ghana's universal healthcare policies are contributing to the reduction of psychosocial stress associated with healthcare inequalities, and therefore, lowering the risk of NCDs in neighborhoods with good healthcare. Nonetheless, the country faces a huge gap in health infrastructure and technology for diagnosing and treating NCDs, which has implications for addressing the rising NCD burden in the country in the coming years (Laar et al., 2019; Ministry of Health, 2018).

The study findings on individual level NCD risk factors are aligned with the literature and situated within the neighborhood risk factors revealed in this study. For instance, the finding that wealthy individuals have higher risk of multiple NCDs is tied to the effect of structural income inequalities found at the neighborhood level. It also corroborates the position of previous studies. In a study by Kuuire and colleagues, they found that people who had lived their childhood and adulthood in urban neighborhoods in Ghana had higher risks for developing obesity (a known risk for many NCDs) (Kuuire et al., 2019). Tenkorang and Kuuire's nationally representative study in the country found higher odds of cardiovascular diseases among wealthy individuals and given the strength of their analyses, concluded that social gradient and NCD risk were inversely related (Tenkorang & Kuuire, 2016). Similarly, Osman et al. noted higher risk of hypertension among wealthy women in rural localities in Sudan (Osman et al., 2019). Furthermore, age appeared as a strong predictor of NCDs, which corroborates the literature on how physiological and functional decline in old age increase the risk of chronic health conditions (Kämpfen et al., 2018; Pelzom et al., 2017; World Health Organization, 2019). Particularly in deprived contexts such as Ghana, there is also evidence to the effect that poor support systems for old age (e.g., resources, infrastructure, and health services), and changing diets into unhealthy fast-foods as found in this study are contributing to early NCDs (Oyekale, 2019; Rossier et al., 2014; Shrivastava et al., 2016).

While physical activity is confirmed as a predictor of NCD multimorbidity, individuals who rated their physical activeness as moderate (fair) had half the risk of those who reported they were physically active for multiple NCDs. Further analysis (results not reported here) revealed that the majority of 'moderately active' individuals were from low socioeconomic category (poor wealth quintile), who, in the study context, have active normal work life and everyday routines. On the other hand, those who reported being active might have consciously taken to exercising either because of poor health condition(s) or as a strategy of adding physical activity to a sedentary lifestyle. These happen to be the two main drivers of exercising among people in high socioeconomic standing in the study context. Thus, although this second group of individuals could be maintaining active lifestyles, it is possible their normal sedentary working and poor health conditions could have increased their risk for multiple NCDs. By far, physical activity status have been applied in its binary form (active or inactive) (Pelzom et al., 2017; Riha et al., 2014). Such conceptualization masks important nuances in how multiple levels of physical activity contribute to NCD risk, as found in this study. Moreover, higher NCD risk for females and individual with poor SRH were confirmed in our study (Riha et al., 2014).

Despite the robustness of the study analysis and findings, it comes with some limitations worth noting. First, lifestyle factors are self-reported and therefore are prone to bias. In particular, bias may arise as individuals attempt avoiding social stigma associated in reporting alcohol use or consumption of some local meals linked to deprivation. Nonetheless, the data collection process addressed these potential data response biases by using experienced enumerators who assured confidentiality and guaranteed the comfort of respondents during the study. Second the cross-sectional nature of the data makes adjusting for temporariness difficult. For this reason, casual relationships have not been estimated. Finally, capturing more nuanced dimensions of study variables in respect of the frequency and quantity of tobacco and alcohol use could have further enriched the analysis. Similarly, inclusion of social capital which is known to cushion against health risk may have been useful in further analysis of deprivation levels of risk of NCDs (Moore & Kawachi, 2017).

Conclusion

The analysis of neighborhood effects on non-communicable diseases (NCDs) in this study enriches our understanding of how contextual complexities combine with individual factors to influence the rising burden of chronic diseases in Ghana and similar context in SSA. As actions on NCDs take root in the implementation of the Sustainable Development Goals, understanding contextual dynamics have become increasingly relevant to support local level policies and programs. In fact, the WHO independent high-level commission on NCDs identified six priority areas for delivering action on NCDs in member countries, four of which are supported by our study findings (Nishtar et al., 2018). These include: (1) design and implement policies and programs that meet local unique needs, (2) redirect health systems for the promotion and control of NCDs, (3) increase collaboration and regulate interventions of civil societies, academia, government institutions, and communities (neighborhoods), and (4) improve reporting and accountability on NCDs.

Firstly, the revelation of structural income inequality as a critical NCD risk factor in this study reaffirms the negative consequence of unbalanced growth on health and well-being, but in this context, wealthy individuals are rather more vulnerable. Widening income inequality unfortunately has become a significant character of the recent modest economic growth recorded in Ghana, with impacts apparent at national level and in neighborhoods (UNDP, 2019). In this context, pro-physiological and behavioral change policies and programs should be complemented with an agenda for inclusive development, targeted at reducing structural inequalities and creating buffers for vulnerable groups such as the elderly population. At least Ghana and other countries in SSA have an opportunity to forestall the experiences of high-income countries by reorienting their economic development path to create wealth and prosperity for all. Secondly, health systems strengthening is critical for addressing NCDs in Ghana and similar contexts. For starters, universal health insurance and expansion of healthcare services in neighborhoods (e.g., CHPS) can reduce psychosocial stress over poor healthcare access. Policies for training of health personnel should emphasize NCD prevention and control, while more investment into public education for the prevention of NCDs, and in technology, health infrastructure, and health system integration for early diagnosis, treatment, and control/management should be reprioritized. Thirdly, programmatic synergy across government agencies, academia, civil society, and local communities will streamline actions on NCDs, leverage resources/capacities, and improve tracking of progress and emerging complexities in disease multimorbidity. The neglect of data collection and reporting on NCDs in Ghana and similar contexts affects effective policymaking, planning, and implementation of strategies to reduce NCD risks.

Acknowledgments We acknowledge funding from Canada Research Chairs Fund, the Social Sciences and Humanities Research Council of Canada, file no. 430-2018-00460 and the University of Toronto—Mississauga Research and Scholarly Activity Fund (RSAF).

References

Adhikari, B., Pokharel, S., & Mishra, S. R. (2019). Shrinking urban Greenspace and the rise in non-communicable diseases in South Asia: An urgent need for an advocacy. *Frontiers in Sustainable Cities, 1*, 5.

Adjaye-Gbewonyo, K., & Vaughan, M. (2019). Reframing NCDs? An analysis of current debates. *Global Health Action, 12*(1), 1641043.

Agyei-Mensah, S., & de-Graft Aikins, A. (2010). Epidemiological transition and the double burden of disease in Accra, Ghana. *Journal of Urban Health, 87*(5), 879–897.

Akyeampong, K. (2009). Revisiting free compulsory universal basic education (FCUBE) in Ghana. *Comparative Education, 45*(2), 175–195.

Anfaara, F. W., Atuoye, K. N., Mkandawire, P., & Luginaah, I. (2018). Factors associated with voluntary testing for HBV in the Upper West Region of Ghana. *Health & Place, 54*, 85–91.

Antabe, R., Atuoye, K. N., Sano, Y., Kuuire, V. Z., Galaa, S. Z., & Luginaah, I. (2019). Health insurance enrolment in the Upper West Region of Ghana: Does food security matter? *International Journal of Health Planning and Management, 34*, e1621–e1632.

Aryeetey, G. C., Nonvignon, J., Amissah, C., Buckle, G., & Aikins, M. (2016). The effect of the National Health Insurance Scheme (NHIS) on health service delivery in mission facilities in Ghana: A retrospective study. *Globalization and Health, 12*(32), 1–9.

Beaglehole, R., Bonita, R., Horton, R., Adams, C., Alleyne, G., Asaria, P., et al. (2011). Priority actions for the non-communicable disease crisis. *The Lancet, 377*(9775), 1438–1447.

Bennett, J. E., Stevens, G. A., Mathers, C. D., Bonita, R., Rehm, J., Kruk, M. E., et al. (2018). NCD countdown 2030: Worldwide trends in non-communicable disease mortality and progress towards Sustainable Development Goal target 3.4. *The Lancet, 392*(10152), 1072–1088.

Bygbjerg, I. C. (2012). Double burden of noncommunicable and infectious diseases in developing countries. *Science, 337*(6101), 1499–1501.

Cummins, S., Curtis, S., Diez-Roux, A. V., & Macintyre, S. (2007). Understanding and representing "place" in health research: A relational approach. *Social Science and Medicine, 65*(9), 1825–1838. https://doi.org/10.1016/j.socscimed.2007.05.036

Curtis, S. (2016). In A. Williams & S. Elliot (Eds.), *Space, place and mental health* (2nd ed.). Routledge.

Ezzati, M., Vander Hoorn, S., Lawes, C. M. M., Leach, R., James, W. P. T., Lopez, A. D., et al. (2005). Rethinking the "diseases of affluence" paradigm: Global patterns of nutritional risks in relation to economic development. *PLoS Medicine, 2*(5), e133.

Frenk, J., & Gomez-Dantes, O. (2011). The triple burden: Disease in developing nations. *Harvard International Review, 33*(3), 36–40.

Ghana Statistical Service. (2012). *2010 Population and housing census. Ghana Statistical Service*. Accra, Ghana. https://doi.org/10.1371/journal.pone.0104053.

Ghana Statistical Service. (2014). *2010 Population and housing census report: Urbanisation*. Accra, Ghana.

Ghana Statistical Service. (2018). *Ghana living standards survey (Round 7)*. Accra, Ghana.

Ghana Statistical Service. (2020). *Projected population by age and sex, 260 districts 2020*. Accra, Ghana.

Goldstein, H., Browne, W., & Rasbash, J. (2002). Partitioning variation in multilevel models. In *Understanding statistics: Statistical issues in psychology, education, and the social sciences* (Vol. 1, issue 4, pp. 223–231). Lawrence Erlbaum Associates.

Haregu, T. N., Wekesah, F. M., Mohamed, S. F., Mutua, M. K., Asiki, G., & Kyobutungi, C. (2018). Patterns of non-communicable disease and injury risk factors in Kenyan adult population: A cluster analysis. *BMC Public Health, 18*(1225), 54–113.

Hu, F., Hu, B., Chen, R., Ma, Y., Niu, L., Qin, X., & Hu, Z. (2014). A systematic review of social capital and chronic non-communicable diseases. *Bioscience Trends, 8*(6), 290–296.

Hurst, J. R., Dickhaus, J., Maulik, P. K., Miranda, J. J., Pastakia, S. D., Soriano, J. B., et al. (2018). Global alliance for chronic disease researchers' statement on multimorbidity. *Lancet Global Health, 6*(12), e1270–e1271.

Iyer, H. S., James, P., Valeri, L., Bajunirwe, F., Nankya-Mutyoba, J., Njelekela, M., et al. (2021). Neighborhood greenness and burden of non-communicable diseases in Sub-Saharan Africa: A multi-country cross-sectional study. *Environmental Research, 196*, 110397.

Kämpfen, F., Wijemunige, N., & Evangelista, B. (2018). *Aging, non-communicable diseases, and old-age disability in low-and middle-income countries: A challenge for global health* (Vol. 63, pp. 1011–1012). International Journal of Public Health.

Kawachi, I., Kennedy, B. P., & Glass, R. (1999). Social capital and self-reported health: A contextual analysis. *American Journal of Public Health, 89*(8), 1187–1193. https://doi.org/10.2105/AJPH.89.8.1187

Kearns, R. A. (1993). Place and health: Towards a reformed medical geography*. *The Professional Geographer, 45*(2), 139–147.

Kuuire, V. Z., Bisung, E., & Were, J. M. (2019) Examining the connection between residential histories and obesity among Ghanaians: evidence from a national survey. *Journal of Public Health, 27*(5), 569–579. https://doi.org/10.1007/s10389-018-0983-8

Laar, A. K., Adler, A. J., Kotoh, A. M., Legido-Quigley, H., Lange, I. L., Perel, P., & Lamptey, P. (2019). Health system challenges to hypertension and related non-communicable diseases prevention and treatment: Perspectives from Ghanaian stakeholders. *BMC Health Services Research, 19*(1), 1–13.

Lumagbas, L. B., Coleman, H. L. S., Bunders, J., Pariente, A., Belonje, A., & de Cock Buning, T. (2018). Non-communicable diseases in Indian slums: Re-framing the social determinants of health. *Global Health Action, 11*(1), 1438840.

Lynch, J., & Smith, G. D. (2005). A life course approach to chronic disease epidemiology. *Annual Review of Public Health, 26*, 1–35.

Marmot, M., & Bell, R. (2019). Social determinants and non-communicable diseases: Time for integrated action. *BMJ, 364*, l251.

Marmot, M., Friel, S., Bell, R., Houweling, T. A. J., Taylor, S., & Commission on Social Determinants of Health. (2008). Closing the gap in a generation: Health equity through action on the social determinants of health. *The Lancet, 372*(9650), 1661–1669.

Mikkelsen, B., Williams, J., Rakovac, I., Wickramasinghe, K., Hennis, A., Shin, H.-R., et al. (2019). Life course approach to prevention and control of non-communicable diseases. *BMJ, 364*, l257.

Ministry of Health. (2018). *Holistic Assessment of 2017 health sector program of work.* Accra, Ghana.

Moore, S., & Kawachi, I. (2017). Twenty years of social capital and health research: A glossary. *Journal of Epidemiology Community Health, 71*(5), 513–517.

Morgan, S. A., Eyles, C., Roderick, P. J., Adongo, P. B., & Hill, A. G. (2019). Women living with multi-morbidity in the Greater Accra Region of Ghana: A qualitative study guided by the cumulative complexity model. *Journal of Biosocial Science, 51*(4), 562–577.

Munodawafa, D. (2019). Gaborone call to action on noncommunicable diseases to achieve Sustainable Development Goals# 3. *International Journal of Noncommunicable Diseases, 4*(2), 56–57.

Nishtar, S., Niinistö, S., Sirisena, M., Vázquez, T., Skvortsova, V., Rubinstein, A., et al. (2018). Time to deliver: Report of the WHO Independent High-Level Commission on NCDs. *The Lancet, 392*(10143), 245–252.

Nugent, R., Bertram, M. Y., Jan, S., Niessen, L. W., Sassi, F., Jamison, D. T., et al. (2018). Investing in non-communicable disease prevention and management to advance the Sustainable Development Goals. *The Lancet, 391*(10134), 2029–2035.

Osman, S., Costanian, C., Annan, N. B., Fouad, F. M., Jaffa, M., & Sibai, A. M. (2019). Urbanization and socioeconomic disparities in hypertension among older adult women in Sudan. *Annals of Global Health, 85*(1), 29.

Oyekale, A. S. (2019). Effect of obesity and other risk factors on hypertension among women of reproductive age in Ghana: An instrumental variable probit model. *International Journal of Environmental Research and Public Health, 16*(23), 4699.

Pelzom, D., Isaakidis, P., Oo, M. M., Gurung, M. S., & Yangchen, P. (2017). Alarming prevalence and clustering of modifiable noncommunicable disease risk factors among adults in Bhutan: A nationwide cross-sectional community survey. *BMC Public Health, 17*(975), 1–11.

Puoane, T. R., Tsolekile, L. P., Caldbick, S., Igumbor, E. U., Meghnath, K., & Sanders, D. (2012). Chronic non-communicable diseases in South Africa: Progress and challenges. *South African Health Review, 2012/2013*(1), 115–126.

Riha, J., Karabarinde, A., Ssenyomo, G., Allender, S., Asiki, G., Kamali, A., et al. (2014). Urbanicity and lifestyle risk factors for cardiometabolic diseases in rural Uganda: A cross-sectional study. *PLoS Medicine, 11*(7), e1001683.

Rosenberg, M. W. (1998). Medical or health geography? Populations, peoples and places. *International Journal of Population Geography, 4*(3), 211.

Rossier, C., Soura, A. B., Duthé, G., & Findley, S. (2014). Non-communicable disease mortality and risk factors in formal and informal neighborhoods, Ouagadougou, Burkina Faso: Evidence from a health and demographic surveillance system. *PLoS One, 9*(12), e113780.

Santosa, A., Wall, S., Fottrell, E., Högberg, U., & Byass, P. (2014). The development and experience of epidemiological transition theory over four decades: a systematic review. *Global Health Action, 7*(1), 23574.

Shrivastava, S. R., Shrivastava, P. S., & Ramasamy, J. (2016). World Health Organization advocates for a healthy diet for all: Global perspective. *Journal of Research in Medical Sciences, 21*, 44.

Spiteri, J., & von Brockdorff, P. (2019). Economic development and health outcomes: Evidence from cardiovascular disease mortality in Europe. *Social Science & Medicine, 224*, 37–44.

Stanifer, J. W., Egger, J. R., Turner, E. L., Thielman, N., & Patel, U. D. (2016). Neighborhood clustering of non-communicable diseases: Results from a community-based study in Northern Tanzania. *BMC Public Health, 16*(226), 1–10.

Tenkorang, E. Y., & Kuuire, V. Z. (2016). Noncommunicable diseases in Ghana: Does the theory of social gradient in health hold? *Health Education & Behavior, 43*(1_suppl), 25S–36S.

UNDP. (2019). *Human development report 2019: Beyond income, beyond averages, beyond today*. New York.

World Bank. (2015). *Rising through cities in Ghana: Ghana urbanization review*. Washington, D.C.

World Bank. (2020). *Life expectancy at birth, total (years)*.

World Health Organization. (2013). *Global action plan for the prevention and control of non-communicable diseases*. Geneva, Switzerland.

World Health Organization. (2018a). *Global health estimates 2016: Disease burden by cause, age, sex, by country and by region, 2000–2016*. Geneva, Switzerland.

World Health Organization. (2018b). *Projections of mortality and causes of death, 2017–2060*. Geneva, Switzerland.

World Health Organization. (2019). *Preparing for an ageing population*. WHO Regional Office for the Western Pacific.

Chapter 3
Correlates of Hypertension Among Women in Ghana: Evidence from the Women's Health Survey

Obinna C. Ezeagbor and Eric Y. Tenkorang

Introduction

Non-communicable diseases (NCDs) include cancer, hypertension, diabetes mellitus, chronic respiratory diseases, and other chronic conditions or illnesses caused by non-disease causing agents (Nigel, 2001a). Globally, NCDs are believed to be the leading cause of morbidity and mortality contributing to about 60% of deaths and 46% of the total burden of disease (Murray & Lopez, 1996). The World Health Organization (WHO) estimated that NCDs accounted for over half of the 57 million deaths in 2008 with most of these linked to cardiovascular diseases, such as hypertension, diabetes, angina, and other chronic respiratory diseases (WHO, 2011). The steady rise in the incidence of NCDs is a major challenge on the global health agenda and poses significant concerns for policymakers. While developed countries contribute substantially to NCDs and NCD-related deaths globally, developing countries are known to bear the largest burden. For instance, the WHO estimates that of the 38 million NCD-related deaths that occur annually around the world, 28 million are in low and middle-income countries (WHO, 2011).

Hypertension is one of such NCDs with increasing prevalence around the globe. Hypertension poses unique complications, which could result in the damage of critical or vital organs in the body and subsequently lead to mortality. It is estimated that about 1 billion people in the world suffer from hypertension, causing roughly 7.1 million deaths annually (Bruntland, 2002). In the past, developed countries were believed to share the highest burden of hypertension cases. However, studies have shown that the incidence of hypertension in some developing countries, including those in Africa, has increased markedly over the last few years, especially in urban

O. C. Ezeagbor · E. Y. Tenkorang (✉)
Department of Sociology, Memorial University of Newfoundland, St. John's, Canada
e-mail: eytenkorang@mun.ca

J. A. Braimah et al. (eds.), *Health Geography in Sub-Saharan Africa*, Global Perspectives on Health Geography, https://doi.org/10.1007/978-3-031-37565-1_3

areas (Nigel, 2001b). And with prevalence ranging from 30 to 40%, the WHO declared hypertension a public health threat in most African countries (Addo et al., 2007).

Ghana, like other countries in Africa, continues to witness an increasing prevalence of hypertension (Aikins et al., 2012; Ministry of Health, 2011). For instance, the Ghana Health Service identified hypertension as the leading cause of death in the country, with prevalence ranging from 19 to 48% among adults (Bosu, 2010; Ghana Health Service, 2007). The sudden rise in NCDs, including hypertension, poses serious challenges to Ghana's epidemiologic transition and reflects the changing socio-economic circumstances in the country.

The recent classification of Ghana as a lower middle-income country and the discovery of oil in commercial quantities are known to have strengthened the Ghanaian economy (Tenkorang et al., 2015a, b). Tenkorang and Kuuire (2016) argued that these developments in Ghana's economy are strongly associated with the country's demographic and health transitions. It is important to mention, however, that, unlike the western industrialized countries where economic development led to significant improvements in morbidity and mortality, especially for communicable diseases, this has not been the same for Ghana. Currently, Ghana is experiencing a 'double burden of disease' where common communicable diseases such as malaria, typhoid, and diphtheria exist simultaneously with NCDs and contribute substantially to morbidity and mortality. Although progress has been made in reducing mortality from communicable diseases, these conditions still account for 3 out of 4 premature deaths (IHME, 2015). At the same time, deaths from NCDs have emerged as the leading causes of years lost, accounting for about 25% of total deaths in sub-Saharan Africa (IHME, 2015).

While hypertension affects both men and women in Ghana, it is documented that women are more affected (Bate et al., 2009, Sen & Ostlin, 2008). Yet, there is very little academic research published on the prevalence of hypertension among women in Ghana and sub-Saharan Africa more broadly and the factors that influence their risk of becoming hypertensive. In fact, the United Nations and other international bodies identify the lack of studies in this area as a neglected dimension of women's health (IOM, 2010). This is equally regrettable given women's significant role in the household, especially in the African setting. Specifically, this study will examine the contributions of socio-economic, lifestyle, and psychosocial factors to women's risks of becoming hypertensive in Ghana.

Hypertension Among Ghanaian Women

The effects of hypertension are present in both men and women globally, but there are gender differences in prevalence. A report on Nigeria indicates that compared to 38.6% of males, 41.2% of females suffered from hypertension (WHO, 2011). In Ghana, the prevalence of hypertension among women was estimated at 7.5% (Nyarko, 2016). But this is only self-reported data that are subject to memory and recall bias. Also, given that hypertension is a 'silent killer,' many women living with it may not be aware of their condition (WHO, 2011). Biometric or clinical data that show more accurate estimates indicate a prevalence of 44.1% (Tenkorang et al., 2015a, b).

The rise in the prevalence of hypertension among women in Ghana can be attributed to several factors, including unhealthy lifestyle, rapid urbanization, population aging, global marketing of tobacco, and food (Smith & Mensah, 2003). There are documented scholarly linkages between sedentary lifestyle and hypertension risk (Bosu, 2010; Miszkurka et al., 2012; Setel, 2003). Owusu-Sekyere et al. (2013) found that the prevalence of hypertension was high among females compared to males due mainly to women living sedentary lifestyles. Ismail et al. (2013) further indicate that sedentary workers, especially market women, bankers, and office workers, have a high prevalence of hypertension and must be considered an occupational risk group. A direct positive relationship exists between living a sedentary lifestyle and physical inactivity. It is thus not surprising that the extant literature finds physical inactivity a strong and important cause of hypertension (Addo et al., 2006). A study conducted in Accra shows that sedentary workers such as market women have a high risk of being hypertensive as a result of the high level of physical inactivity associated with such occupations (Konadu et al., 2016).

The lack of physical activity has left quite a substantial proportion of Ghanaian women overweight and obese. A recent study in the Greater Accra region estimated the crude obesity rates for Ghanaian women as 20.2% compared to 4.6% for their male counterparts (Amoah et al., 2002). Also, data from the Ghana Demographic and Health Survey show a 2.5-fold increase in obesity rates among women, from 10% in 1993 to 25.3% in 2003 (Ghana Statistical Service, 2004). Meanwhile, obesity has been documented as a significant correlate of cardiovascular ailments such as hypertension in several populations (Donkor et al., 2015; Kotsis et al., 2010), although theoretical links between the two are not clearly known. Similar to physical inactivity and obesity, tobacco use has been identified as one of the many risk factors for hypertension (Tonstad & Johnston, 2006). Although there is a significant reduction in the prevalence of tobacco use in industrialized countries (Ng et al., 2014), evidence shows a growing epidemic of tobacco use in developing countries. In sub-Saharan Africa, including Ghana, smoking is common among males, with a prevalence ranging between 20% and 60% (Pampel, 2005). However, Addo et al. (2009) found a smoking prevalence of 0.3% among Ghanaian women. While the link between tobacco use and hypertension is unclear, some argue that the presence of nicotine in tobacco can lead to elevated blood pressure (Tuomiletho et al., 1982).

Socio-economic factors such as income, education, and occupation have also been identified as important determinants of the risks of living with hypertension among women (Donkor et al., 2015). It is important to highlight, however, that the relationship between socio-economic factors and the risk of developing NCDs including hypertension has largely remained inconclusive. For instance, while some research shows that wealthier and highly educated women are significantly more likely to live with NCDs, including hypertension, other documented evidence indicates a lower risk of hypertension among wealthy and educated individuals. Higher risks of hypertension among the educated and wealthy in sub-Saharan Africa have often been traced to lifestyle factors (Murphy et al., 2013; Negin et al., 2011; Tenkorang & Kuuire, 2016). For women with limited education and economic opportunities, poverty could be the underlying risk factor for hypertension (Opie & Seedat, 2005; Seedat, 2007). We contribute to the debates linking socio-economic variables such as income and education to the risks of living with hypertension in Ghana and sub-Saharan Africa in general.

Data and Methods

The data employed for analysis comes from Wave II of the Women's Health Study of Accra (WHSA II), conducted from 2008 to 2009 by the Institute of Statistical, Social and Economic Research (ISSER), Ghana, and the Department of Global Health and Population, Harvard School of Public Health. This study was conducted primarily to obtain information on the linkages between health and wealth at the household level. The WHSA II was designed as a community-based study to quantify the magnitude of the burden of communicable and non-communicable diseases among a representative sample of adult women aged 18 and older residing in the Accra Metropolitan Area. The study is built on a previous survey from Wave I of WHSA II, which answered questions regarding the nature of health conditions in the general population. Accra was the selected study site because it shares many of the characteristics of other rapidly growing urban populations in West Africa (Hill et al., 2007).

The WHSA II is a multistage representative sample of 3200 women aged 18 years and over who were primary participants from Wave I of the study (WHSA-1), conducted in 2003. The study area consists of 11 sub-metropolitan areas and one Municipal Assembly of the larger city of Accra with 1731 Enumeration Areas (EAs). At the first stage of sampling, 200 EAs were selected as primary sampling units using probability proportional to size from 1731 EAs, with an estimated population of 2.2 million people living in roughly 364,000 households (Ghana Statistical Service, 2005). All women aged 18 years and older were listed and an extensive mapping was carried out in the selected EAs to ensure representation. The listed women formed the frame for the second stage of selection. Only one woman per household was selected and interviewed at the second stage of sampling.

Participants of WHSA I were selected by a two-stage cluster probability sample stratified by socio-economic status based on the 2000 Ghanaian census data. When a participant from WHSA I could not be located, a woman of similar age, socio-economic status, and geographic location as the initial participant in WHSA I replaced the woman in the sample. A total of 995 replacement women were interviewed for a total of 2814 WHSA II participants interviewed between October 2008 and June 2009. Older women were over-sampled to provide enough elderly cases for analysis. The sample for this study was restricted to 2505 women aged 18 years and above whose information on hypertension was available. Ethical clearance was obtained from the WHO for this survey. Thus, all participants provided informed and written consent.

Measures

The dependent variable used for analysis is dichotomous, indicating whether respondents are hypertensive or not. Hypertension status was determined by taking a repeated blood pressure reading or measurements of the respondents. The biometric measurements were done by trained individuals, with the first measurement taken thirty minutes after the start of the interview to avoid an invalid blood pressure reading. The second and third readings were taken subsequently after the interview had been completed, with a fourth measurement taken in the case of any discrepancy. The study made use of the conventional measurement of high blood pressure recommended by the WHO. Individual respondents were classified as hypertensive if the average of their first three systolic blood pressure measurements was equal to or greater than 140 mmHg and/or the average of the first three diastolic blood pressure measurements was equal to or greater than 90 mmHg (≥ 140/ ≥90 mm Hg) (WHO, 2016). Following the WHO protocol, women who were pregnant and all those with missing data were excluded.

Independent variables are grouped into socio-economic, psychosocial, and lifestyle factors. Socio-economic predictors include respondents' education (coded 0 = no education, 1 = primary education, 2 = middle school/JSS, 3 = secondary/SSS, 4 = higher), a derived income variable, created from a series of questions tapping the wealth status of households (coded 0 = poorest, 1 = poorer, 2 = middle, 3 = richer, 4 = richest), and the main occupation of participants (coded 0 = unemployed, 1 = government employee, 2 = private business, 3 = self-employed, 4 = other).

Five lifestyle factors (see Table 3.1 for description) were included in the analysis, those related to nutrition, physical activity, body mass index, smoking, and drinking. Some socio-demographic and cultural variables are used as controls. These include the age of respondents measured in complete years, ethnicity (0 = Akan, 1 = Ga Adangbe, 2 = Ewe, 3 = Northern languages, 4 = other ethnic groups), marital status (0 = never married, 1 = currently married, 2 = separated/divorced/widowed), and religious affiliation (0 = Christian, 1 = Muslim, 2 = no religion).

Data Analysis

Univariate, bivariate, and multivariate analysis were conducted on the data. Binary logistic regression was employed, given the dichotomous nature of the dependent variable. In the bivariate analysis, we examined the gross effect of socio-economic, lifestyle, psychosocial, and some demographic/cultural variables on the likelihood

Table 3.1 Lifestyle and psychosocial variables

Variable	Description	Measurement
Nutrition		
Vegetable consumption	Derived from two questions that asked respondents *whether they consumed fresh/salad vegetables such as lettuce, carrots, and tomatoes;* and *whether they consumed any darky leafy vegetables such as kontomire, aleefu, ayoyo, kale, and cassava leaves in the past 24 h*	0 = 'no vegetables', 1 = 'at least one vegetable', 2 = '2 or more vegetables'
Fruits consumption	Derived from three questions that asked respondents *whether they consumed fruits such as pawpaw, apple, mango, orange, or pineapple in the past 24 h; whether they drank 100% fruit juice such as orange and grape fruit juice in the past 24 h;* and *whether they drank other sweetened beverage such as regular soda, tea with sugar, or fruit flavored drinks in the past 24 h*	0 = 'no fruits' for respondents who did not consume any fruit, 1 = 'low fruits' for respondents who consumed only one fruit, 2 = 'high fruits' for respondents who consumed two or more fruits
Protein consumption	Derived from seven questions that asked respondents *whether they ate yogurt in the past 24 h; whether they ate red meat such as beef, pork, goat, grasscutter, or bush meat in the past 24 h; whether they ate chicken in the past 24 h; whether they ate fish in the past 24 h; whether they ate eggs in the past 24 h; whether they ate agushie in the past 24 h; and whether they ate groundnuts in the past 24 h*	0 = 'low protein' for respondents who had at most 1 protein source, 1 = 'moderate protein' for respondents who had between 2 and 3 protein sources, 2 = 'high protein' for respondents who had at least 4 protein sources
Fat consumption	Derived from five questions that asked respondents *whether they ate fried chicken, fish, or vegetables in the past 24 h; whether they used butter or margarine on bread or on foods in the past 24 h; whether they ate groundnuts, cashews, seeds, or other nuts in the past 24 h; whether they ate fast food snacks in the past 24 h;* and *what type of fat they use or is usually used at their homes to cook food*	0 = 'no fat' for respondents who did not eat any food containing fat, 1 = 'low fat' for respondents who ate only one food containing fat, 2 = 'high fat' for respondents who ate 2 or more foods containing fat
Physical activity	A composite variable from four questions that asked respondents *whether their work involve moderately intense activities, like brisk walking or carrying light loads; whether their work involve vigorous activity, like heavy lifting, digging, or heavy manual work; whether they do vigorous activities like playing basketball, running, competitive swimming, or playing ball in their free time;* and *whether they do moderate intensity activities like brisk walking, gentle swimming, or exercising in their free time*	0 = 'no exercise' for respondents who did not engage in any physical activity, 1 = 'moderate/low exercise' for respondents who engaged in only one kind of physical activity, 2 = 'high/intensive exercise' for respondents who engaged in 2 or more kinds of physical activity

(continued)

Table 3.1 (continued)

Variable	Description	Measurement
Smoking	*Currently smoke cigarettes*	0 = No, 1 = Yes
Drinking	*Have ever consumed a drink that contains alcohol such as beer, wine, gin, bitters, palm wine, or schnapps*	0 = No, 1 = Yes
Body mass index	Created from anthropometric measures (height and weight of respondents)	0 = underweight, 1 = normal, 2 = overweight, 3 = obese
Psychosocial	An additive scale derived from six questions that asked respondents *how often they felt depressed in the last month, how often they felt hopeless in the last month, how often they felt restless or fidgety in the last month, how often they felt so depressed that nothing could cheer them up in the last month, how often they felt that everything was an effort in the last month* and *how often they felt worthless in the last* month	A continuous variable

of becoming hypertensive among the respondents. In the multivariate models, we examined the net effects of socio-economic predictors (education, wealth status, occupation), demographic variables (age of respondent, sex, marital status), lifestyle (physical exercise, vegetable consumption, fruits consumption, alcohol consumption, protein consumption, fat consumption), and psychosocial factors on the likelihood of becoming hypertensive among the respondents. The multivariate models are sequential such that the first model includes socio-economic variables, the second model added the lifestyle variables, and the third model added the psychosocial factors. Using STATA 13, a cluster variable was imposed on the models to adjust the standard errors hence producing statistically robust parameter estimates (Tenkorang & Owusu, 2010). In all the models, we controlled for ethnicity, marital status, religious affiliation, and age.

Results

Table 3.2 shows the distribution of selected dependent and independent variables and describes the sample used for the study. The majority of the respondents are married and Christians. Results also indicate that the majority of women in the sample are self-employed and have at least a middle school/JSS education. Furthermore, most respondents do not consume vegetables, fruits, and foods that contain fat. In addition, very few of the respondents identified as smokers (0.8%) while quite a large number of the respondents consumed alcohol (46.9%).

Bivariate results are shown in Table 3.3. Some socio-economic factors are significant predictors of hypertension among women in Ghana. Results show that highly educated and employed women are significantly less likely to be hypertensive than their uneducated and unemployed counterparts. Some lifestyle variables are significantly associated with the risk of becoming hypertensive among Ghanaian women. Obese women are significantly more likely to be hypertensive than women who are underweight. Similarly, women that engage in high/intensive exercise are significantly less likely to be hypertensive than women who engage in no exercise. Compared to those who did not consume alcohol at all, women who consumed alcohol are significantly more likely to be hypertensive. Surprisingly, women who consumed fat were significantly less likely to be hypertensive compared to those that did not consume fat at all. Also, the odds of becoming hypertensive were associated with old age and being separated/divorced/widowed.

Multivariate results are presented in Table 3.4. Three models were computed for the respondents. Model 1 examines the effects of socio-economic factors on the likelihood of women being hypertensive, controlling for the demographic characteristics of respondents. Model 2 adds lifestyle variables, and model 3 includes psychosocial factors. Results indicate that women with higher education are significantly less likely to be hypertensive than those without education. Similarly, women who are self-employed are significantly less likely to be hypertensive compared to those who are unemployed. The risk for self-employed women was accentuated when both lifestyle and psychosocial factors were controlled for in models 2 and 3. The magnitude of the risks for women with higher education widened and the coefficient became highly significant when psychosocial factors were controlled for. Results also show that women who are obese are significantly more likely to be hypertensive than women who are underweight. Separated/divorced/widowed women are significantly more likely to be hypertensive compared to married women. Similarly, older women are more likely to be hypertensive compared to younger women.

Discussions

The WHO has described hypertension as a 'silent killer and a global public health crisis. Currently, it is estimated that over nine million people live with hypertension globally, the majority of whom reside in low- and middle-income countries, including those in sub-Saharan Africa (WHO, 2013). While prevalent among all demographic groups, hypertension is most common among women (Bates et al., 2009; Sen & Ostlin, 2008). Yet few studies explore women's vulnerability and the factors influencing their risk of becoming hypertensive.

Results show high prevalence of hypertension among Ghanaian women sampled in this study, which compares and is consistent with studies found elsewhere (Nyarko, 2016; Tenkorang et al., 2015a, b). The findings corroborate others that argue that hypertension among women is often underestimated, feeding into the misperception that women are at lower risk than men (Gudmundsdottir et al., 2012). Given that women are unique and may be affected by hypertension differently from their male counterparts, the paper explored if women's risk factors may also be unique. Results show that Ghanaian women with higher education are significantly

Table 3.2 Univariate distribution of selected dependent and independent variables

Variables	% (N = 2505)
Diagnosed with hypertension?	
Not hypertensive	61.9
Hypertensive	38.2
Socio-economic variables	
Education	
None	21.4
Primary	12.1
Middle School/JSS	39.6
Secondary/SSS	17.0
Higher	10.0
Wealth status	
Poorer	20.0
Poorest	20.0
Middle	20.0
Richer	20.0
Richest	20.0
Occupation	
Unemployed	27.3
Government employee	4.9
Private business	8.3
Self-employed	51.5
Other	8.0
Lifestyle factors	
Body mass index	
Under weight	3.7
Normal	30.9
Overweight	28.0
Obese	37.5
Physical exercise	
No exercise	39.9
Moderate/low exercise	48.9
High/intensive exercise	11.2
Vegetable consumption	
No vegetables	64.4
At least 1 vegetable	27.8
2 or more vegetables	7.8
Fruits consumption	
No fruits	61.5
Low fruits	34.7
High fruits	3.9
Smoking	
No	99.2
Yes	0.8

(continued)

Table 3.2 (continued)

Variables	% (N = 2505)
Alcohol consumption	
No	53.1
Yes	46.9
Protein consumption	
Low	52.4
Moderate	40.9
High	6.8
Fat consumption	
No	61.1
Low	28.9
High	10.0
Psychosocial factors	
Psychosocial (mean)	2.9
Controls	
Ethnicity	
Akan	32.3
Ga/Adangbe	41.3
Ewe	13.5
Northern languages	5.3
Other	7.7
Marital status	
Never married	18.6
Currently married	50.4
Separated/divorced/widowed	31.0
Religious affiliation	
Christian	82.8
Muslim	12.6
No religion	2.1
Other	2.5
Age	
18–24	27.9
25–34	21.9
35–54	25.5
55 and above	24.8

Table 3.3 Bivariate analysis of hypertension among women in Ghana

Predictor variables	OR(SE)
Socio-economic variables	
Education	
None	1.00
Primary	0.476 (0.07)***
Middle School/JSS	0.412 (0.04)***
Secondary/SSS	0.272 (0.04)***
Higher	0.214 (0.03)***
Wealth Status	
Poorer	1.00
Poorest	0.787 (0.10)*
Middle	0.734 (0.10)*
Richer	0.858 (0.11)
Richest	0.815 (0.11)
Occupation	
Unemployed	1.00
Government employee	0.514 (0.10)***
Private business	0.235 (0.05)***
Self-employed	0.530 (0.05)***
Other	0.445 (0.09)***
Lifestyle factors	
Body mass index	
Under weight	1.00
Normal	0.619 (0.14)*
Overweight	1.163 (0.28)
Obese	1.935 (0.43)**
Physical exercise	
No exercise	1.00
Moderate/low exercise	0.735 (0.06)***
High/intensive exercise	0.628 (0.09)***
Vegetable consumption	
No vegetables	1.00
At least 1 vegetable	0.991 (0.09)
2 or more vegetables	0.878 (0.13)
Fruits consumption	
No fruits	1.00
Low fruits	0.911 (0.08)
High fruits	0.692 (0.15)
Smoking	
No	1.00
Yes	0.926 (0.39)
Alcohol consumption	
No	1.00

(continued)

Table 3.3 (continued)

Predictor variables	OR(SE)
Yes	1.189 (0.10)*
Protein consumption	
Low	1.00
Moderate	0.936 (0.08)
High	0.774 (0.15)
Fat consumption	
No	1.00
Low	0.767 (0.07)**
High	0.655 (0.08)***
Psychosocial factors	
Psychosocial	1.018 (0.01)
Controls	
Ethnicity	
Akan	1.00
Ga/Adangbe	1.488 (0.14)***
Ewe	1.150 (0.15)
Northern languages	0.940 (0.16)
Other	1.438 (0.22)*
Marital Status	
Never married	1.00
Currently married	3.163 (0.46)***
Separated/divorced/widowed	7.030 (0.96)***
Religious affiliation	
Christian	1.00
Muslim	1.130 (0.13)
No religion	0.933 (0.27)
Other	0.913 (0.23)
Age	
18–24	1.00
25–34	2.312 (0.31)***
35–54	5.689 (0.72)***
55 and above	9.437 (1.27)***
N	2505
Log-likelihood (ll)	−1719.6

Note: *$p < 0.1$, **$p < 0.05$; ***$p < 0.01$, *the first category of each categorical variable was set as reference, standard errors (SE) are in parenthesis, OR odds ratio*

less likely to be hypertensive. The finding is largely consistent with the theory of social gradient in health (Kostenuik & Dickinson, 2003), which suggests that individuals with lower SES (socio-economic status) are often burdened with diseases compared with those with higher SES (Tenkorang & Kuuire, 2016). It corroborates several other studies that indicate the protective effects of formal education on health outcomes (Bloom, 2007; Mirowsky & Ross, 2003).

Table 3.4 Multivariate analysis of hypertension among women in Ghana

	Model 1	Model 2	Model 3
	OR(SE)	OR(SE)	OR(SE)
Socio-economic variables			
Education			
None	1.00	1.00	1.00
Primary	0.711 (0.12)*	0.676 (0.12)*	0.671 (0.12)*
Middle School/JSS	0.694 (0.08)**	0.661 (0.08)***	0.658 (0.08)***
Secondary/SSS	0.609 (0.10)**	0.575 (0.10)**	0.569 (0.10)**
Higher	0.404 (0.08)***	0.403 (0.09)***	0.408 (0.09)***
Wealth status			
Poorer	1.00	1.00	1.00
Poorest	1.074 (0.15)	0.968 (0.13)	0.928 (0.13)
Middle	1.017 (0.14)	0.876 (0.13)	0.875 (0.13)
Richer	1.307 (0.21)	1.171 (0.20)	1.169 (0.19)
Richest	1.188 (0.20)	1.045 (0.18)	1.023 (0.18)
Occupation			
Unemployed	1.00	1.00	1.00
Government employee	1.186 (0.28)	1.070 (0.28)	1.159 (0.31)
Private business	0.671 (0.15)	0.659 (0.16)	0.670 (0.16)
Self-employed	0.763 (0.09)*	0.723 (0.09)**	0.757 (0.09)*
Other	0.742 (0.14)	0.699 (0.15)	0.741 (0.16)
Lifestyle factors			
Body mass index			
Under weight		1.00	1.00
Normal		0.751 (0.18)	0.776 (0.19)
Overweight		1.220 (0.32)	1.276 (0.35)
Obese		1.876 (0.48)*	1.997 (0.52)**
Physical exercise			
No exercise		1.00	1.00
Moderate/low exercise		1.022 (0.11)	1.036 (0.12)
High/intensive exercise		0.922 (0.16)	0.943 (0.17)
Vegetable consumption			
No vegetables		1.00	1.00
At least 1 vegetable		0.865 (0.09)	0.857 (0.10)
2 or more vegetables		0.838 (0.16)	0.840 (0.16)
Fruits consumption			
No fruits		1.00	1.00
Low fruits		0.971 (0.10)	0.960 (0.10)
High fruits		0.908 (0.24)	0.895 (0.24)
Smoking			
No		1.00	1.00
Yes		0.693(0.41)	0.683(0.41)

(continued)

Table 3.4 (continued)

	Model 1	Model 2	Model 3
	OR(SE)	OR(SE)	OR(SE)
Alcohol consumption			
No		1.00	1.00
Yes		1.124 (0.12)	1.119 (0.12)
Protein consumption			
Low		1.00	1.00
Moderate		1.154 (0.12)	1.170 (0.12)
High		1.427 (0.34)	1.470 (0.35)
Fat consumption			
No		1.00	1.00
Low		0.944 (0.10)	0.960 (0.11)
High		0.925 (0.15)	0.905 (0.15)
Psychosocial factors			
Psychosocial			1.018 (0.01)
Controls			
Ethnicity			
Akan	1.00	1.00	1.00
Ga/Adangbe	1.037 (0.11)	0.981 (0.11)	0.989 (0.11)
Ewe	1.010 (0.15)	0.895 (0.13)	0.879 (0.13)
Northern languages	0.780 (0.21)	0.707 (0.21)	0.698 (0.21)
Other	1.141 (0.29)	1.200 (0.31)	1.158 (0.30)
Marital status			
Never married	1.00	1.00	1.00
Currently married	1.348 (0.24)	1.115 (0.21)	1.075 (0.20)
Separated/divorced/widowed	1.666 (0.32)**	1.457 (0.28)	1.422 (0.28)
Religious affiliation			
Christian	1.00	1.00	1.00
Muslim	1.223 (0.30)	1.093 (0.30)	1.102 (0.30)
No religion	0.671 (0.22)	0.745 (0.25)	0.729 (0.24)
Other	0.947 (0.24)	1.144 (0.31)	1.112 (0.31)
Age			
18–24	1.00	1.00	1.00
25–34	1.884 (0.29)***	1.846 (0.30)***	1.858 (0.30)***
35–54	4.132 (0.62)***	3.929 (0.63)***	3.928 (0.64)***
55 and above	5.379 (0.97)***	4.997 (0.94)***	5.120 (0.98)***
N	2505	2505	2505
Log-likelihood (ll)	−1545.1	−1433.8	−1402.9

Note: *p < 0.1, **p<. 05; ***p<. 01, *the first category of each categorical variable was set as reference, standard errors (SE) are in parenthesis, OR = odds ratio*

Studies that demonstrate education's protective effects on the risks of developing hypertension often refer to access to resources, including health information, as empowering the highly educated to engage in health-promoting behaviors compared to those with no education who may lack such resources (Mackenbach et al., 2008; Steenland et al., 2002). It is important to note, however, that the effects of education on the risk of hypertension remain inconclusive. While some studies find no significant relationship between the two variables (see Kaplan et al., 2010; Thawornchaisit et al., 2013), others find rather that higher education is a risk factor for hypertension in some sub-Saharan African countries (see Addo et al., 2006; Donkor et al., 2015; Nyarko, 2016; Tenkorang et al., 2015a, b).

It is also observed that the risk of hypertension is significantly lower for self-employed women. Previous studies that explored the effects of occupational categories on the risks of hypertension found that the self-employed tend to be highly educated and wealthy (Tenkorang et al., 2015a, b). It is possible, therefore, that the resources that accrue to them resulting from their higher education and wealth may serve as a buffer against developing raised blood pressure.

It was not surprising to find that lifestyle factors such as body mass index significantly affect the risk of becoming hypertensive for women in Ghana. In this regard, the finding that Ghanaian women who are obese are more likely to be hypertensive is instructive and consistent with other studies that identify obesity as one of the major determinants of hypertension among women in sub-Saharan Africa (Commodore-Mensah et al., 2014; Keetile et al., 2015; Ng et al., 2014; Njelekela et al., 2009). This finding may in part be attributed to unhealthy dietary habits and the limited physical activity among Ghanaians and especially among women (Donkor et al., 2015). Some studies show that compared to their male counterparts, Ghanaian females are significantly less likely to engage in physical activities (Tenkorang & Kuuire, 2016).

Unlike previous studies in sub-Saharan Africa (Gebreselassie & Padyab, 2014; Hendriks et al., 2012), lifestyle factors such as smoking, alcohol consumption, protein consumption, fat consumption, physical inactivity, and vegetable consumption did not have a significant effect on the risks of hypertension.

There was no significant relationship between psychosocial factors and the risk of hypertension among Ghanaian women (Jones et al., 1996; Shinn et al., 2001). Although not the focus of this paper, it is worth mentioning and discussing the significance of some control variables. For instance, the finding that older women aged 55 and above are significantly more likely to be hypertensive compared to younger women aged 18–24 is consistent with evidence elsewhere (Bosu, 2010; Pinto, 2007; Ugwuja et al., 2015). It is not surprising to find that separated/divorced/widowed women are more likely to be hypertensive and that single women had a lower likelihood of being hypertensive than married women. The risks of separated/divorced/widowed women may reflect the emotional challenges women face after the loss of a husband or in the event of divorce (see Tenkorang et al., 2015a, b).

The results suggest that policymakers improve the socio-economic conditions of women in Ghana as these have implications for their morbidity risks, including hypertension. Our study also revealed that the majority of women in the sample were overweight and obese. Meanwhile, very few engaged in physical activity and had limited intake of nutritious food including fruits and vegetables. It must be noted that being overweight was significantly associated with the risks of hypertension. Thus, the results suggest that it is important to inform and educate women about the importance of eating healthy and engaging in physical activities. Behavioral change interventions that are gender sensitive and focus on changing the lifestyle of Ghanaian women must be implemented to deal with the high prevalence of hypertension among this important demographic group.

Despite the findings, some limitations are noticeable. First, as data are cross-sectional, we limit the interpretation of the results to associations and desist from making 'causal' connections between predictor and outcome variables. We are equally cautious about generalizing findings to the whole country, given that data were collected in only one of the sixteen administrative regions, the Greater Accra Region. Notwithstanding, this study is one of the few, if not the foremost, to examine the risk factors of hypertension among women in Ghana and has important policy implications.

References

Addo, J., Amoah, A. G., & Koram, K. A. (2006). The changing patterns of hypertension in Ghana: A study of four rural communities in the Ga District. *Ethnicity & Disease, 16*(4), 894–899.

Addo, J., Smeeth, L., & Leon, D. A. (2007). Hypertension in sub-Saharan Africa: A systematic review. *Hypertension, 50*(6), 1012–1018.

Addo, J., Smeeth, L., & Leon, D. A. (2009). Smoking patterns in Ghanaian civil servants: Changes over three decades. *International Journal of Environmental Research and Public Health, 6*(1), 200–208.

Aikins, A., Addo, J., Ofei, F., Bosu, W. K., & Agyemang, C. (2012). Ghana's burden of chronic non-communicable diseases: Future directions in research, practice and policy. *Ghana Medical Journal, 46*(2), 1–3.

Amoah, A. G., Owusu, S. K., & Adjei, S. (2002). Diabetes in Ghana: A community based prevalence study in Greater Accra. *Diabetes Research and Clinical Practice, 56*(3), 197–205.

Bates, L. M., Hankivsky, O., & Springer, K. W. (2009). Gender and health inequities: A comment on the final report of the WHO commission on the social determinants of health. *Social Science & Medicine, 69*(7), 1002–1004.

Bloom, D. E. (2007). *Education, health, and development*. American Academy of Arts and Sciences.

Bosu, W. K. (2010). Epidemic of hypertension in Ghana: A systematic review. *BMC Public Health, 10*(1), 1.

Brundtland, G. H. (2002). Reducing risks to health, promoting healthy life. *JAMA, 288*(16), 1974–1974.

Commodore-Mensah, Y., Samuel, L. J., Dennison-Himmelfarb, C. R., & Agyemang, C. (2014). Hypertension and overweight/obesity in Ghanaians and Nigerians living in West Africa and industrialized countries: A systematic review. *Journal of Hypertension, 32*(3), 464–472.

Donkor, C., Edusei, A. K., Mensah, K. A., Nkoom, B., Okyere, P., Appiah-Brempong, E., & Adjei, R. O. (2015). Prevalence of hypertension and obesity among women in reproductive age in the Ashaiman Municipality in the Greater Accra Region of Ghana. *Prevalence, 5*(2), 89–96.

Gebreselassie, K., & Padyab, M. (2014). Epidemiology of hypertension stages in two countries in sub-Sahara Africa: Factors associated with hypertension stages. *International Journal of Hypertension, 2015*, 959256.

Ghana Health Service. (2007). *Hypertension is the number one killer disease in Ghana.* Accra, Ghana.

Ghana Statistical Service. (2004). *Ghana demographic health survey 2003.* Accra: GSS.

Ghana Statistical Service. (2005). *Population data analysis report: Socioeconomic and demographic trends.* Accra, Ghana Statistical Service.

Gudmundsdottir, H., Høieggen, A., Stenehjem, A., Waldum, B., & Os, I. (2012). Hypertension in women: Latest findings and clinical implications. *Therapeutic Advances in Chronic Disease, 3*(3), 137–146.

Hendriks, M. E., Wit, F. W., Roos, M. T., Brewster, L. M., Akande, T. M., de Beer, I. H., et al. (2012). Hypertension in sub-Saharan Africa: Cross-sectional surveys in four rural and urban communities. *PLoS One, 7*(3), e32638.

Hill, A. G., Darko, R., Seffah, J., Adanu, R. M., Anarfi, J. K., & Duda, R. B. (2007). Health of urban Ghanaian women as identified by the Women's Health Study of Accra. *International Journal of Gynecology & Obstetrics, 99*(2), 150–156.

Institute for Health Metrics and Evaluation. (2015). *Global burden of disease collaborative network. Global burden of disease study 2013 (GBD 2013) incidence, prevalence, and years lived with disability 1990–2013.* Seattle, USA.

IOM (Institute of Medicine). (2010). *Women's Health Research: Progress, Pitfalls, and Promise.* Washington, DC: The National Academies Press.

Ismail, I. M., Kulkarni, A. G., Kamble, S. V., Borker, S. A., Rekha, R., & Amruth, M. (2013). Prevalence of hypertension and its risk factors among bank employees of SulliaTaluk, Karnataka. *Sahel Medical Journal, 16*(4), 139–143.

Jones-Webb, R., Jacobs, D. R., Flack, J. M., & Liu, K. (1996). Relationships between depressive symptoms, anxiety, alcohol consumption, and blood pressure: Results from the CARDIA study. *Alcoholism: Clinical and Experimental Research, 20*(3), 420–427.

Kaplan, M. S., Huguet, N., Feeny, D. H., & McFarland, B. H. (2010). Self-reported hypertension prevalence and income among older adults in Canada and the United States. *Social Science & Medicine, 70*(6), 844–849.

Keetile, M., Navaneetham, K., & Letamo, G. (2015). Patterns and determinants of hypertension in Botswana. *Journal of Public Health, 23*(5), 311–318.

Konadu, P. A., Yekple, C., Adu, I., Adu, J., & Abokyi, E. (2016). Sedentary work and incidence of hypertension among market women in Accra-Ghana: A case study of Makola market women. *Research Journal of Public Health, 2*(2), 1–16.

Kosteniuk, J. G., & Dickinson, H. D. (2003). Tracing the social gradient in the health of Canadians: Primary and secondary determinants. *Social Science & Medicine, 57*(2), 263–276.

Kotsis, V., Stabouli, S., Papakatsika, S., Rizos, Z., & Parati, G. (2010). Mechanisms of obesity-induced hypertension. *Hypertension Research, 33*(5), 386–393.

Mackenbach, J. P., Stirbu, I., Roskam, A. J. R., Schaap, M. M., Menvielle, G., Leinsalu, M., & Kunst, A. E. (2008). Socioeconomic inequalities in health in 22 European countries. *New England Journal of Medicine, 358*(23), 2468–2481.

Ministry of Health. (2011). *National Policy for the prevention and control of chronic non-communicable diseases in Ghana.* Accra, Ghana.

Mirowsky, J., & Ross, C. E. (2003). *Education, social status, and health.* Transaction Publishers.

Miszkurka, M., Haddad, S., Langlois, É. V., Freeman, E. E., Kouanda, S., & Zunzunegui, M. V. (2012). Heavy burden of non-communicable diseases at early age and gender disparities in an adult population of Burkina Faso: World Health Survey. *BMC Public Health, 12*(1), 24–34.

Murphy, G. A. V., Asiki, G., Ekoru, K., Nsubuga, R. N., Nakiyingi-Miiro, J., Young, E. H., et al. (2013). Sociodemographic distribution of non-communicable disease risk factors in rural Uganda: A cross-sectional study. *International Journal of Epidemiology, 42*, 1740–1753.

Murray, C. J. L., & Lopez, A. D. (1996). *The global burden of disease: A comprehensive assessment of mortality and disability from diseases, injuries and risk factors in 1990 and projected to 2020.* Harvard School of Public Health.

Negin, J., Cumming, R., de Stewart Ramirez, R., Abimbola, S., & Sachs, S. E. (2011). Risk factors for non-communicable diseases among older adults in rural Africa. *Tropical Medicine and International Health, 16*, 640–646.

Ng, M., Fleming, T., Robinson, M., Thomson, B., Graetz, N., Margono, C., et al. (2014). Global, regional, and national prevalence of overweight and obesity in children and adults during 1980–2013: A systematic analysis for the Global Burden of Disease Study 2013. *The Lancet, 384*(9945), 766–781.

Nigel, U. (2001a, March 1). *Taking poverty to heart non communicable diseases and the poor*. University Of Newcastle, Department Of Diabetes and Epidemiology & Public Health.

Nigel, U. (2001b). Non-communicable diseases in sub-Saharan Africa: Where do they feature in the health research agenda? Policy and Practice. *Bulletin of the World Health Organization, 79*(10), 947–953.

Njelekela, M. A., Mpembeni, R., Muhihi, A., Mligiliche, N. L., Spiegelman, D., Hertzmark, E., et al. (2009). Gender-related differences in the prevalence of cardiovascular disease risk factors and their correlates in urban Tanzania. *BMC Cardiovascular Disorders, 9*(1), 30.

Nyarko, S. H. (2016). Prevalence and sociodemographic determinants of hypertension history among women in reproductive age in Ghana. *International Journal of Hypertension, 2016*(3), 1–6.

Opie, L. H., & Seedat, Y. K. (2005). Hypertension in sub-Saharan African populations. *Circulation, 112*(23), 3562–3568.

Owusu-Sekyere, E., Bonyah, E & Ossei, L. (2013). Spatial modeling of hypertension disease in the kumasi metropolitan area of Ghana. *International Journal of Statistics and Applications, 3*(4), 132–140.

Pampel, F. C. (2005). Patterns of tobacco use in the early epidemic stages: Malawi and Zambia, 2000-2002. *American Journal of Public Health, 95*(6), 1009–1015.

Pinto, E. (2007). Blood pressure and ageing. *Postgraduate Medical Journal, 83*(976), 109–114. https://doi.org/10.1136/pgmj.2006.048371

Seedat, Y. (2007). Impact of poverty on hypertension and cardiovascular disease in sub-Saharan Africa. *Cardiovascular Journal of Africa, 18*(5), 316–320.

Sen, G., & Ostlin, P. (2008). Gender inequity in health: why it exists and how we can change it. *Global Public Health, 3*(Suppl 1), 1–12.

Setel, P. W. (2003). Non-communicable diseases, political economy, and culture in Africa: Anthropological applications in an emerging pandemic. *Ethnicity & Disease, 13*(2 Suppl), S149.

Shinn, E. H., Poston, W. S. C., Kimball, K. T., Jeor, S. T. S., & Foreyt, J. P. (2001). Blood pressure and symptoms of depression and anxiety: A prospective study. *American Journal of Hypertension, 14*(7), 660–664.

Smith, S. M., & Mensah, G. A. (2003). Population aging and implications for epidemic cardiovascular disease in Sub-Saharan Africa. *Ethnicity and Disease, 13*(Supp 2), S77–S80.

Steenland, K., Henley, J., & Thun, M. (2002). All-cause and cause-specific death rates by educational status for two million people in two American Cancer Society cohorts, 1959–1996. *American Journal of Epidemiology, 156*(1), 11–21.

Tenkorang, E. Y., & Kuuire, V. Z. (2016). Non-communicable diseases in Ghana does the theory of social gradient in health hold? *Health Education & Behavior, 43*(1 suppl), 25S–36S.

Tenkorang, E. Y., & Owusu, G. (2010). Correlates of HIV/AIDS testing in Ghana: Some evidence from the demographic and health survey. *AIDS Care, 22*(3), 296–307.

Tenkorang, E. Y., Kuuire, V., Luginaah, I., & Banchani, E. (2015a). Examining risk of actors for hypertension in Ghana: Evidence from the study on global ageing and adult health. *Global Health Promotion, 24*(1), 14–26.

Tenkorang, E., Sano, Y., Sedziafa, A. P., Kuuire, V., & Banchani, E. (2015b). Validity of self-report data in hypertension research; Findings from the study on global ageing and adult health. *Journal of Clinical Hypertension, 17*(12), 977–984.

Thawornchaisit, P., de Looze, F., Reid, C. M., Seubsman, S. A., Sleigh, A. C., Chokhanapitak, J., et al. (2013). Health risk factors and the incidence of hypertension: 4-year prospective findings from a national cohort of 60 569 Thai Open University students. *BMJ Open, 3*(6), e002826.

Tonstad, S., & Johnston, J. A. (2006). Cardiovascular risks associated with smoking: A review for clinicians. *European Journal of Cardiovascular Prevention & Rehabilitation, 13*(4), 507–514.

Tuomilehto, J., Elo, J., Nissmen, A. (1982). Smoking among patients with malignant hypertension. *BMJ, 1*, 1086.

Ugwuja, E. I., Ezenkwa, U. S., Nwibo, A. N., Ogbanshi, M., Idoko, O., & Nnabu, R. (2015). Prevalence and determinants of hypertension in an agrarian rural community in Southeast Nigeria. *Annals of Medical and Health Sciences Research, 5*(1), 45–49.

WHO. (2013). *Global strategy on diet, physical activity and health*. WHO Fact Sheet. World Health Organization. Available at http://www.int/dietphysicalactivity/

World Health Organization. (2011). *Hypertension*. World Health Organization. http://www.searo.who.int/entity/noncommunicable_diseases/media/non_communicable_diseases_hypertension_fs.pdf

World Health Organization. (2016). *Raised blood pressure: Situation and trends*. World Health Organization. http://www.who.int/gho/ncd/risk_factors/blood_pressure_prevalence_text/en/

Chapter 4
Exploring Coping Strategies of Persons with Mental Illness in Ghana: A Synthesis of the Qualitative Literature

Joseph Asumah Braimah, Ebenezer Dassah, Elijah Bisung, and Mark W. Rosenberg

Introduction

Mental disorders have emerged as a significant public health concern over the past decades (World Health Organization, 2013). Globally, in 2015, over 300 million (4.4% of the population) and 264 million (3.6% of the population) people were estimated to be living with depression and anxiety disorders, respectively (World Health Organization, 2017). In the case of Ghana, as of 2015, the proportion of the population with depressive disorders and anxiety disorders were estimated at 4.2% and 2.8%, respectively, translating into 8.7% and 3.0% of all Years Lived with Disability (World Health Organization, 2017). Considering the central role that good mental health plays in achieving overall health and well-being, there is an urgent need for relevant stakeholders to (re)direct efforts toward preventing mental disorders and promoting mental well-being.

J. A. Braimah (✉)
Department of Health and Society, University of Toronto Scarborough, Toronto, ON, Canada
e-mail: j.braimah@utoronto.ca

E. Dassah
School of Public Health, Kwame Nkrumah University of Science and Technology, Kumasi, Ghana
e-mail: edassah10@knust.edu.gh

E. Bisung
School of Kinesiology and Health Studies, Queen's University, Kingston, ON, Canada
e-mail: elijah.bisung@queensu.ca

M. W. Rosenberg
Department of Geography and Planning, Queen's University, Kingston, ON, Canada
e-mail: mark.rosenberg@queensu.ca

J. A. Braimah et al. (eds.), *Health Geography in Sub-Saharan Africa*, Global Perspectives on Health Geography, https://doi.org/10.1007/978-3-031-37565-1_4

Mental health, as defined by the World Health Organization, is "a state of well-being whereby individuals recognize their abilities, are able to cope with the normal stresses of life, work productively and fruitfully, and make a contribution to their communities" (World Health Organization, 2003, p. 7). Mental disorders, on the other hand, are "mental health problems severe enough to reach the threshold of a diagnosis within the internationally-agreed psychiatric classification systems" (Organisation for Economic Co-operation and Development, 2010, p. 6). The Ghana Mental Health Act 846 defines mental disorder as a "condition of the mind in which there is a clinically significant disturbance of mental or behaviour functioning associated with distress or interference of daily life and manifesting as disturbance of speech, perception, mood, thought, volition, orientation or other cognitive functions to such degree as to be considered pathological but excludes social deviance without personal dysfunction" (Government of Ghana, 2012, p. 39). While different kinds of mental disorders exist, depression and anxiety disorders have been reported to be the most prevalent (World Health Organization, 2017). Mental illness, on the other hand, is a broader concept encompassing both mental disorders and mild or moderate mental health problems that do not reach the threshold of diagnosis (Organisation for Economic Co-operation and Development, 2010).

Similar to most health conditions, mental illness presents a unique and hefty burden on individuals and society at large, especially in low- and middle-income countries where healthcare systems are poorly developed. Specifically, mental disorders have been linked to morbidity and mortality (Ohrnberger et al., 2017; Prince et al., 2007; Read et al., 2009). They have also profoundly contributed to poverty (Lund et al., 2010; World Health Organization, 2003). Other negative consequences of mental disorders are violence, trauma, low productivity, depletion of household resources, stigma, and discrimination (Barke et al., 2011; Gyamfi et al., 2018; Lund et al., 2010; Tawiah et al., 2015).

In response to the high rates of mental disorders and associated burdens, the WHO and other international organizations have over the years embarked on some key global agendas to promote mental well-being. For instance, the Mental Health Action Plan 2013–2020 was developed "to promote mental well-being, prevent mental disorders, provide care, enhance recovery, promote human rights and reduce the mortality, morbidity, and disability for persons with mental disorders" (World Health Organization, 2013, p. 9). This action plan also provides guidance for country-level action plans. Similarly, the Sustainable Development Goals (Goal 3.4) underscores the need to promote mental health and well-being among people (United Nations, 2015). At the national level, the Ghana Mental Health Act was passed in 2012, among other initiatives, to promote a comprehensive, modern, and community-based mental health system as well as protect the rights of persons with mental disorders (Government of Ghana, 2012; Walker & Osei, 2017).

Despite these initiatives, evidence reveals a wide mental healthcare access gap in Ghana. Several individual and structural factors account for the poor mental healthcare system in Ghana. The lack of commitment to implement existing mental health policies, inadequate funding, and the lack of coordination within the national healthcare delivery system undermine the provision of mental healthcare services

(Ghana Health Service, 2017; Roberts et al., 2014; Walker & Osei, 2017). At the individual level, poverty, stigma, discrimination, lack of education, and inherent cultural practices and beliefs have been identified as barriers to mental health service uptake in the country (Gyamfi et al., 2018).

In Ghana, there are some relatively recent reviews that have focused on mental health issues, including access to mental health services (Badu et al., 2018) and research priorities on mental health (see Quansah & Karikari, 2016; Read & Doku, 2012). To date, however, reviews have not focused on the coping strategies adopted by persons with mental illness to deal with their situation in Ghana. Therefore, this review aims to close this knowledge gap by identifying and synthesizing existing qualitative empirical studies that focus on how persons with mental illness cope with their conditions in Ghana. As the global community continues to promote mental well-being, synthesizing coping strategies could help generate policy-relevant interventions and clinical guidelines for addressing mental illness in Ghana.

Conceptual Framework

Coping usually describes individuals' dispositional characteristics in dealing with stressors over time and in various situations consistently (Coppens et al., 2010). Folkman & Lazarus (1980, 1984) conceptualized coping as an attempt to manage internal and/or external stressors and demands that are appraised by individuals as taxing or exceeding their existing resources. The authors further categorized coping into either problem focused or emotion focused. Problem focused denotes the individual's attempts to solve the issue at stake or seek support to manage the stressor. Emotion focused, on the other hand, reflects the individual's attempt to turn away from the source of stress by escaping the problems, engaging in self-accusation, and denying reality (Lazarus & Folkman, 1984). Skinner et al. (2003) expand upon these coping categories to develop a taxonomy of coping that is comprised of five core domains: problem solving, support seeking, avoidance, distraction, and positive cognitive restructuring. We adopt this coping typology proposed by Skinner et al. (2003) to synthesize existing empirical evidence on the coping strategies adopted by people with mental illness in Ghana. Table 4.1 presents definitions of the five domains.

Table 4.1 Definition of coping domains by Skinner et al. (2003)

Domain	Definition
1. Problem solving	Concerns efforts to engage in specific problem solving steps
2. Support seeking	Involves efforts to obtain assistance (such as emotional and instrumental support) from others
3. Avoidance	Relates to efforts to stay away from or not deal with the problem or situation at hand
4. Distraction	Involves engaging oneself in pleasurable activities (such as hobbies, exercising, and substance use) to manage stress
5. Positive cognitive restructuring	Relates to changing one's perspective of a stressful situation in order to see it in a more positive light

Methods

We followed the scoping review methodology presented by Arksey and O'Malley (2005). Scoping reviews typically focus on systematically mapping the current state and breadth of the phenomenon of study and identifying patterns and gaps to inform the development of recommendations for future research (Arksey & O'Malley, 2005; Pham et al., 2014). We searched for relevant articles on mental health and coping strategies in Ghana in the following academic electronic databases: Scopus, Embase, Medline, and PsycINFO. We used key search terms related to coping, mental health, mental disorders, mental illness, and Ghana. Our search was performed in June 2022.

We included articles based on the following criteria: (a) English peer-reviewed articles published between January 2000 and December 2019 to capture the most recent evidence on the topic before the COVID-19 pandemic; (b) primary/empirical articles that focus on mental health coping strategies; and (c) qualitative or mixed methodologies, provided the qualitative portion of the mixed-method study is sufficiently developed and retrievable. We excluded general literature describing the process of care or access to mental healthcare services. We did not include literature reviews, policy papers, commentaries, and short communications. We, however, used them to contextualize the discussion of the findings.

A total of 842 articles were imported into Endnote reference management software. We also conducted a hand search of relevant mental health journals and reviewed the reference lists of relevant articles to identify additional papers. Finally, we used the titles of potentially eligible articles on Google Scholar's features "cited by" and "related articles" to retrieve relevant additional articles. These procedures yielded additional 18 studies. The titles of relevant articles were read and screened by two reviewers and were organized into two Endnote files, one with the included articles and the other with excluded studies. These Endnote files were checked by a third reviewer. Following the removal of duplicates, 731 articles were retained. The titles of these articles were further screened, and this process produced 168 studies. Abstracts and full-text screening were again independently conducted by one reviewer and checked independently by a second reviewer, reducing the number of included studies to 26. Conflicting opinions about the eligibility of any articles were discussed and resolved by all three reviewers. We identified a total of 9 articles that were used for the review. The diagram below details how the articles were selected (see Fig. 4.1).

Two reviewers extracted data from the 9 included studies using Microsoft Excel. Data extraction included information about the author(s) and year of publication, participants' characteristics, study methods, and a summary of the key findings related to coping mechanisms among persons with mental health issues. We defined data as information that appeared in the results/findings section of each article. We also included verbatim quotes from participants in the individual studies and results/findings reported by the authors. Finally, we did not include the discussion and conclusion sections in the synthesis. These procedures have been effectively used in similar qualitative reviews (Houghton et al., 2016; Reilly & Houghton, 2019). The results of the review are presented in two different ways. First, the characteristics of the included articles are presented in a table, followed by the themes of the included studies mapped deductively onto the coping framework by Skinner et al. (2003).

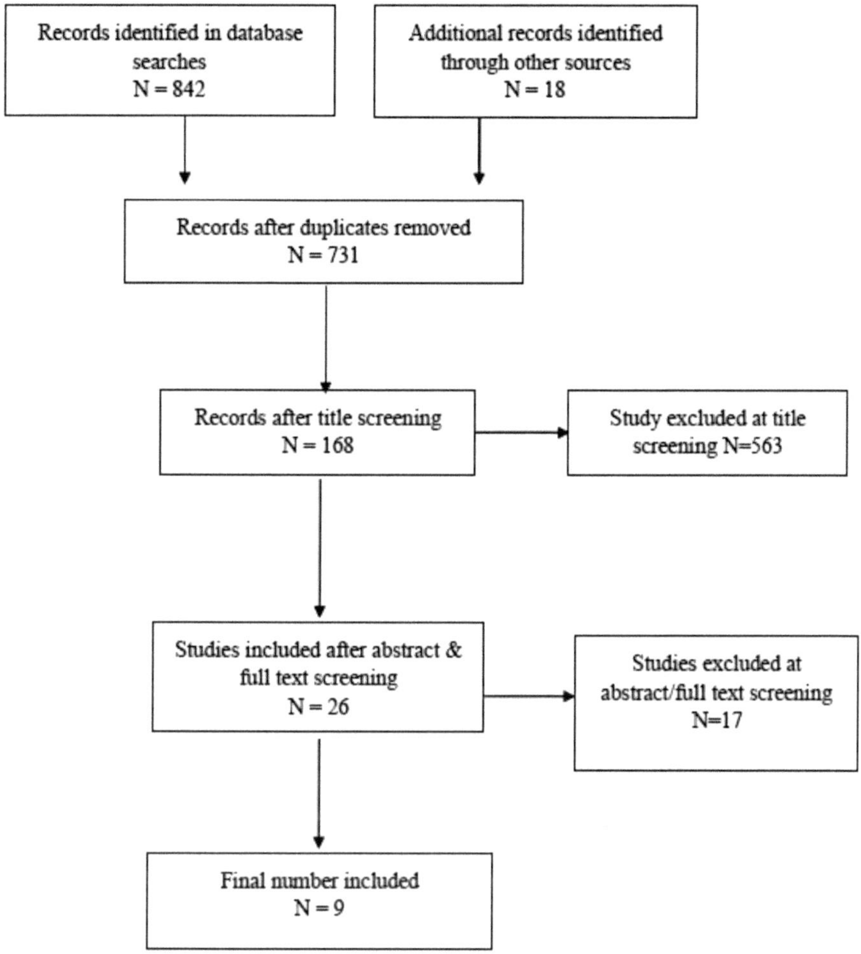

Fig. 4.1 PRISMA flow diagram

Results

Characteristics of the Included Articles

The final set of articles includes 8 qualitative studies and 1 mixed method study. Most of the qualitative studies did not provide information about the specific qualitative design used. Two articles mentioned ethnographic approach (Read, 2012; Read et al., 2009). In-depth and semi-structured interviews, and focus group discussions were the common data collection methods documented in the articles. Two articles additionally employed field observations in the data collection process (Read, 2012; Read et al., 2009). Thematic and framework analysis were the primary data analysis techniques used in most of the articles. The number of persons who participated in the studies varied from 8 to 277. The articles focused mostly on adult populations (i.e., 18 years and above) and included both males and females in their samples. These participants were primarily drawn from persons with mental health issues (e.g., alcoholism, bipolar disorder, epilepsy, and schizophrenia) and their caregivers (e.g., family members), traditional and faith-based healers (e.g., pastors and imams), mainstream mental health practitioners, and policy makers. See Table 4.2 for details of the included articles.

Table 4.2 Characteristics of studies

No.	Author(s) and year of publication	Study participants	Study methods	Key findings related to coping strategies and mental healthcare
1	Ae-Ngibise et al. (2010)	122 different participants (healthcare providers, users of psychiatric services, teachers, police officers, academics, and religious and traditional healers)	*Design*: Qualitative *Data collection*: In-depth interviews and focus group discussions *Data Analysis*: Framework analysis	*Cognitive*: Faith in traditional healers and pastors could address psychological problems and also allay the fears associated with mental health *Problem solving*: Reliance on religious and faith-based healers to solving mental health problems *Support seeking*: Psychosocial support from traditional and faith healers
2	Ae-Ngibise et al. (2015)	75 caregivers of persons with mental issues (33 males and 42 females) aged between 21 and 45 years	*Design*: Qualitative *Data collection*: In-depth interviews and focus group discussions *Data Analysis*: Framework analysis	*Cognitive*: Caregivers used prayers, being hopeful for miracles, and positive about new treatments for mental health *Escape-avoidance*: Participants gave up, only hoping for miracles *Problem solving*: Reliance on prayers in solving mental health problems *Support seeking*: Caregivers relied on faith-based healers for spiritual support

(continued)

Table 4.2 (continued)

No.	Author(s) and year of publication	Study participants	Study methods	Key findings related to coping strategies and mental healthcare
3	Gyamfi et al. (2018)	12 persons with mental health issues (9 males and 3 females) aged between 18 and 50 years	*Design*: Qualitative *Data collection*: Semi-structured interviews *Data Analysis*: Thematic analysis	*Cognitive*: Maintaining relationships with others helped participants remain positive about their lives. They were also optimistic that medical treatment and/or their faith could relieve their symptoms and enable them to recover *Distraction*: Participants recounted watching television and listening to gospel music *Escape-avoidance*: Participants coped through physical or emotional isolation *Problem solving*: Participants relied on religious and faith-based healers' prayers *Support seeking*: Reliance on traditional and faith-based healers for spiritual support
4	Kpobi and Swartz (2018a)	10 pastors/leaders from self-styled Charismatic churches (8 males and 2 females) aged between 31 to 66 years	*Design*: Qualitative *Data collection*: Semi-structured interviews *Data Analysis*: Thematic analysis	*Cognitive*: Participants maintain that faith in God could work through others to use medicine to cure people with mental health issues *Problem solving*: Participants association with God enabled them to provide prayers that helped persons with mental health cope with their situation *Support seeking*: Participants provided spiritual counselling and directions
5	Kpobi and Swartz (2018b)	8 traditional healers (7 males and 1 female) aged between 53 to 73 years	*Design*: Qualitative *Data collection*: Interviews *Data Analysis*: Thematic analysis	*Cognitive*: Focused on gods to heal with culturally specific treatment methods *Support seeking*: Participants provided spiritual support to persons with mental health issues

(continued)

Table 4.2 (continued)

No.	Author(s) and year of publication	Study participants	Study methods	Key findings related to coping strategies and mental healthcare
6	Kpobi et al. (2019)	8 traditional herbalists (6 males and 2 female) aged between 47 to 62 years	*Design: Qualitative* *Data collection*: Interviews *Data Analysis*: Thematic analysis	*Cognitive*: Participants offered prayers to their clients *Problem solving*: Participants used various herbal remedies to treat their clients *Support seeking*: Participants provided spiritual support to their clients
7	Read et al. (2009)	67 different participants — 25 People with mental illness —31 Caregivers —3 Traditional healers —4 Pastors —1 Muslim healer —3 Imams	*Design*: Qualitative ethnography study *Data collection*: Semi-structured interviews, focus group discussions, and observations *Data Analysis*: Grounded theory approach	*Problem solving*: Participants relied on traditional and faith-based healers' prayers *Support seeking*: Persons with mental health received both material and spiritual support from faith-based healers
8	Read (2012)	67 different participants — 25 People with mental illness —31 Caregivers —3 Traditional healers —4 Pastors —1 Muslim healer —3 Imams	*Design*: Qualitative ethnography study *Data collection*: Semi-structured interviews, focus group discussions, and observations *Data Analysis*: Not specified	*Problem solving*: Beatings in prayer camps could solve mental health problems
9	Tawiah et al. (2015)	277 persons with mental health issues (105 males and 172 females)	*Design*: Mixed methods *Data collection*: Interviews, focus groups, and surveys *Data Analysis*: Descriptive and thematic analysis	*Distraction*: Participants engaged in psychological activities and economic ventures that acted as distraction techniques *Problem solving*: Participants resorted to prayer camps and traditional healers *Support seeking*: Persons with mental health received support from family members

Coping Strategies

Problem Solving

Nearly all the articles reported a wide range of problem solving strategies by persons with mental illness. In particular, the use of "conventional" mental health practitioners' medications was common. However, some persons with mental health issues continue to doubt the efficacy of such medications because of their side effects, including weakness and tiredness (Read, 2012). As such, the use of traditional and faith-based healers was highlighted as widespread and thus played a significant role in solving mental health problems (Ae-Ngibise et al., 2010; Gyamfi et al., 2018). Some of the studies further explained that the wide patronage of traditional and faith healers is deeply rooted in culturally constructed models that explain the origins of mental illness (Ae-Ngibise et al., 2010). Given this, some patients patronize traditional healers to "receive spiritual guidance for the correct treatment" (Ae-Ngibise et al., 2010, p. 561). Other religious strategies employed by faith-based healers included prayers, fasting, and the use of anointing oils or holy water (Ae-Ngibise et al., 2010, 2015; Kpobi & Swartz, 2018a; Tawiah et al., 2015). Other controversial problem solving practices found in prayer camps included beatings. *"Beatings were also part of treatment to rid the person of evil spirits which were perceived by both pastors and traditional healers to lie behind much mental illness"* (Read et al., 2009, p. 9). Read et al. (2009) further indicated that beating was regarded to be necessary because it coerced persons with mental illness to confess their wrongdoings, and this, in turn, could solve their mental health problems. Chains and shackles were also used by faith-based practitioners to restrain mental health patients (Read, 2012; Read et al., 2009). Aside from these troubling healing practices, studies described how smoking marijuana and taking antipsychotics acted as problem solving techniques for persons with mental illness (Read, 2012; Tawiah et al., 2015).

Positive Cognitive Restructuring

Most of the papers indicated that religious beliefs and maintaining friends were important coping resources that helped persons with mental illness downplay their conditions and think positively about new treatment regimes and recovery. Specifically, trust in God provides persons with mental illness and their caregivers an avenue to rekindle their hope and think positively about recovery (Ae-Ngibise et al., 2015; Gyamfi et al., 2018). For example, Gyamfi and colleagues found that "*many participants expressed guarded optimism that ... their faith in God would make their symptoms disappear and they would recover*" (Gyamfi et al., 2018, p. 374). Given this, the articles highlighted that many of these individuals resort to prayers offered by spiritual leaders, including pastors and other divine instructors, or actively search for spiritual healing in prayer camps and churches. Further, some persons with mental illness maintained relationships with others, and such relationships enabled them to think positively about their daily lives while downplaying the adverse effects of their mental health problems (Gyamfi et al., 2018).

Support Seeking

Support seeking was a major coping strategy for persons with mental illness in Ghana (Ae-Ngibise et al., 2010, 2015; Gyamfi et al., 2018; Kpobi & Swartz, 2018a; Read et al., 2009; Tawiah et al., 2015). Participants sought support from diverse sources, including family, friends, and religious leaders. For instance, a study indicated that participants turn to traditional and faith-based healers for emotional support: *"Many respondents highlighted that traditional and faith healers are like 'clinical psychologists', providing 'talk therapy', 'counselling' and 'asking questions and providing solutions"* (Ae-Ngibise et al., 2010, p. 561). This support form was noted to be easily accessible and less expensive (Read et al., 2009). Moreover, participants sought biomedical support from trained psychologists to leverage such spiritual support from religious leaders (Kpobi & Swartz, 2018b). The articles further reported instances where participants believed that seeking biomedical support was revealed to them by God. Sometimes, family members only turn to spiritual and biomedical support when they cannot manage aggressive behaviors at home (Read et al., 2009; Tawiah et al., 2015). Besides spiritual support, persons with mental health problems received material support in the form of food, clothing, and accommodation from pastors (Read et al., 2009). An article provided contrasting views about participants seeking support from the religious leader (Read et al., 2009). On the positive side, support was *"a refuge from stigma in the home community"* for persons with mental health problems (Read et al., 2009). On the flip side, it was an avenue for human rights violations, including chaining and starving from food and water (Read et al., 2009).

Escape-Avoidance

Avoidance was reported in two articles (Ae-Ngibise et al., 2015; Gyamfi et al., 2018). Notable avoidance strategies include refusing to think about their illness, denying that mental illness/stress exists, isolation, secrecy, and engaging in wishful thinking (Ae-Ngibise et al., 2015; Gyamfi et al., 2018). Conceding struggles and isolating from others were the most dominant avoidance behaviors. For instance, persons with mental health physically hide *"to avoid discrimination...[and] to avoid others using the spirits against them"* (Gyamfi et al., 2018, p. 373). Similarly, the articles mentioned that persons with mental health issues were sometimes sent to prayer camps to physically and emotionally isolate them from society.

Distraction

Only two articles discussed distraction as a coping resource (Gyamfi et al., 2018; Tawiah et al., 2015). These papers documented how persons with mental health issues engaged in different distractive activities to divert attention from their circumstances. For instance, activities such as reading the bible, watching TV, and taking substances (marijuana) helped these individuals to release stress and derive pleasure. Other activities included going out for meals, animal rearing, and oversleeping (Tawiah et al., 2015). In particular, persons with mental illness mostly resort to religious activities (e.g., praying and reading the bible) to divert attention from other people (Gyamfi et al., 2018). Gyamfi et al. (2018) further noted how individuals resort to watching TV and listening to music as strategies for reducing stigma and discrimination associated with their mental disorders. A participant recounted his strategy as follows:

> Already most of the time when I see (old friends), I try to dodge. I want to be alone so that no one will point fingers at me anymore or say anything that will hurt me again… when it gets too much, I just stay in my room watch television and listen to gospel music (Gyamfi et al., 2018, p. 373).

Others maintained friendships with colleagues to divert attention from their mental health circumstances (Gyamfi et al., 2018). Finally, Tawiah et al. (2015) showed that psychological activities (e.g., smoking marijuana and oversleeping) and economic ventures (e.g., animal rearing) acted as distraction techniques from mental illness. These strategies have been argued to be consoling, motivating, and emotionally satisfying, possibly explaining their acceptance in Ghana and most parts of the African region.

Discussion

The aim of our study is to synthesize existing qualitative evidence on the coping strategies adopted by persons with mental illness in Ghana. We identified several interrelated and overlapping strategies that are used to cope with mental illness in Ghana. Guided by Skinner et al.'s (2003) coping framework, these strategies have been categorized into five broad themes/domains, namely, problem solving, support seeking, distraction, escape/avoidance, and positive cognitive restructuring. Although mental health has gained considerable attention from researchers and practitioners, the literature on coping strategies adopted by persons with mental health issues is largely underdeveloped, particularly in sub-Saharan Africa.

Several problem solving strategies were deployed to manage the stress associated with mental illness. Problem solving strategies involve those steps taken by patients to overcome stress engendered by the disorder (Skinner et al., 2003). Common problem solving strategies employed included faith healing (through the use of various healing aids such as anointing oils, holy water, and fasting) largely involving traditional healers, pastors/priests, and "mallams" (Gyamfi et al., 2018; Kpobi, 2017; Read et al., 2009; Tawiah et al., 2015), seeking biomedical care (Gyamfi et al., 2018; Read & Doku, 2012), and chaining (Read et al., 2009). The centrality of faith healing in the health system is well documented in Ghana and across the African region, largely stemming from the social construction of mental illness as a disease caused by both physical and spiritual factors (Ae-Ngibise et al., 2010; Burns & Tomita, 2015; Doris & Sylvia, 2015). Against this backdrop, individual behaviors and practices tend to be firmly rooted in their religious beliefs, possibly explaining why mental illness is given religious connotations (Opare-Henaku & Utsey, 2017; Salifu Yendork et al., 2016). In addition to the use of spiritual healing practices, religion also provides psychosocial and emotional support for people, which is critical for their mental well-being. Given its central role in the Ghanaian healthcare system, albeit with concerns about the safety and efficacy of their practices, there is the need for collaboration between faith healing and biomedical care as part of efforts to provide comprehensive mental healthcare services.

Mental healthcare is a core component of Ghana's Healthcare System, with few psychiatric institutions established to provide psychiatric in-patient care. The provision of formalized mental healthcare is also reiterated in Ghana's Mental Health Act (Government of Ghana, 2012). Notwithstanding this, accessing formal psychiatric care services remains a challenge, which may explain the prominence of non-conventional psychiatric care (Burns & Tomita, 2015). For this reason, the World Health Organization (WHO) mental health Gap Action Programme (mhGAP) calls for the need to integrate mental health issues into primary healthcare systems in low- and middle-income countries (World Health Organization, 2008). As such, future experimental studies could explore models of integrating mental healthcare into Ghana's primary healthcare system.

Avoidance strategies could partly be attributed to the inherent discrimination and stigma associated with mental illness in Ghana and similar contexts (Barke et al., 2011; Tawiah et al., 2015). For persons with mental illnesses, a condition not heard or seen is less spoken about and hence, the need to avoid public spaces. Beyond patients deciding to engage in avoidance activities by themselves, our findings revealed that social networks encouraged avoidance (Gyamfi et al., 2018). However, there is growing evidence of the relevance of social networks and public spaces in overcoming psychosocial stress, including mental stress (Kawachi & Berkman, 2001; Kushitor et al., 2018). Social networks remain an integral part of Ghanaian society, providing support for the vulnerable and needy. Unsurprisingly, persons with mental illnesses within this context also rely on these networks to cope with stress. The family is reported as the commonest form of support for these persons. Religious leaders were also a common form of support in coping with mental illnesses (Ae-Ngibise et al., 2010; Tawiah et al., 2015). Support from these social networks come in various forms, such as emotional, psychological, and instrumental.

Positive cognitive restructuring also emerged as a strategy to overcome stress in our integrative review. Indeed, changing one's perspective of a stressful situation to view it in a more positive light has the potential to promote positive health outcomes (Seguin et al., 2017; Skinner et al., 2003). This finding corroborates previous studies that suggest that a positive outlook renders people more resilient (Chukwu et al., 2019; Seguin et al., 2017). Thus, some participants invest time to maintain positive relationships with others in order to downplay their mental health challenges (Gyamfi et al., 2018). Others internalized their health conditions as a form of protection against stress. Portraying a positive image of oneself enhances self-confidence, which has considerable implications for one's identity and well-being. Despite the importance of a self-positive image in managing mental health problems, it is imperative to acknowledge that this is contingent on the severity of the illness.

Limitations

This study has some limitations worth acknowledging. First, we only searched four databases which could potentially limit the number of articles available for our analysis. However, these databases are the most common avenues for retrieving relevant articles on the subject. Also, by reviewing the references of included articles and using google searches, we minimized the likelihood of omitting relevant studies. Second, we did not perform a quality assessment of the articles. However, given that all retrieved articles have been peer reviewed, the chances of using a low-quality article is minimized.

Conclusion

This study presents a synthesis of existing qualitative research on the coping strategies adopted by persons with mental illness in Ghana. Despite efforts to provide formal care services to mentally ill persons, driven by the biomedical ideology, our findings reveal a wide array of informal strategies to overcome the stress associated with mental illness. Our findings underscore the need for scaling-up formal mental care services in Ghana and similar contexts to bridge the existing service gap. We also recommend the integration of mental healthcare services into Ghana's primary healthcare system to improve coverage. There is also the need for coordination between the formal and informal sectors for the provision of comprehensive and appropriate mental healthcare services. We anticipate this review will set the stage for conversations on providing context-relevant interventions or care for persons with mental illness. As rural primary healthcare continues to improve in Ghana and sub-Saharan Africa, options for integrating mental healthcare into Primary Health Care need to be at the forefront of policy discussions.

References

Ae-Ngibise, K. A., Cooper, S., Adiibokah, E., & Akpalu, B. (2010). 'Whether you like it or not people with mental problems are going to go to them': A qualitative exploration into the widespread use of traditional and faith healers in the provision of mental health care in Ghana. *International Review of Psychiatry, 22*(6), 558–567. https://doi.org/10.3109/09540261.2010.536149

Ae-Ngibise, K. A., Doku, V. C. K., Asante, K. P., & Owusu-Agyei, S. (2015). The experience of caregivers of people living with serious mental disorders: A study from rural Ghana. *Global Health Action, 8*(1), 26957.

Arksey, H., & O'Malley, L. (2005). Scoping studies: Towards a methodological framework. *International Journal of Social Research Methodology, 8*(1), 19–32.

Badu, E., O'Brien, A. P., & Mitchell, R. (2018). An integrative review of potential enablers and barriers to accessing mental health services in Ghana. *Health Research Policy and Systems, 16*(1), 110.

Barke, A., Nyarko, S., & Klecha, D. (2011). The stigma of mental illness in Southern Ghana: Attitudes of the urban population and patients ' views. *Soc Psychiatry Psychiatr Epidemiol, 46*, 1191–1202. https://doi.org/10.1007/s00127-010-0290-3

Burns, J. K., & Tomita, A. (2015). Traditional and religious healers in the pathway to care for people with mental disorders in Africa: A systematic review and meta-analysis. *Social Psychiatry and Psychiatric Epidemiology, 50*(6), 867–877.

Chukwu, N. E., Okoye, U. O., Onyeneho, N. G., & Okeibunor, J. C. (2019). Coping strategies of families of persons with learning disability in Imo state of Nigeria. *Journal of Health, Population and Nutrition, 38*(1), 9.

Coppens, C. M., de Boer, S. F., & Koolhaas, J. M. (2010). Coping styles and behavioural flexibility: Towards underlying mechanisms. *Philosophical Transactions of the Royal Society B, 365*(1560), 4021–4028.

Doris, F. A., & Sylvia, A. A. (2015). Perceptions and beliefs about mental illness (Schizophrenia) among adults in Zaare community. *Developing Countries Studies, 5*(9), 150–159.

Folkman, S., & Lazarus, R. S. (1980). An analysis of coping in a middle-aged community sample. *Journal of Health and Social Behavior, 21*(3), 219–239.

Folkman, S., & Lazarus, R. S. (1984). *Stress, appraisal, and coping*. Springer.

Ghana Health Service. (2017). *Ghana Health Service 2016 annual report*. Accra. https://doi.org/10.1136/bjo.2010.193169

Government of Ghana. (2012). Mental Health Act, 2012, Pub. L. No. 846, 1, Ghana.

Gyamfi, S., Hegadoren, K., & Park, T. (2018). Individual factors that influence experiences and perceptions of stigma and discrimination towards people with mental illness in Ghana. *International Journal of Mental Health Nursing, 27*, 368–377. https://doi.org/10.1111/inm.12331

Houghton, C., Murphy, K., Brooker, D., & Casey, D. (2016). Healthcare staffs' experiences and perceptions of caring for people with dementia in the acute setting: Qualitative evidence synthesis. *International Journal of Nursing Studies, 61*, 104–116.

Kawachi, I., & Berkman, L. F. (2001). Social ties and mental health. *Journal of Urban Health, 78*(3), 458–467.

Kpobi, L. (2017). "I' m here because of christ and worshipping god …" factors influencing religious switching among Ghanaian charismatic/Neo-pentecostal Christians. *Archive for the Psychology of Religion, 39*, 295–311. https://doi.org/10.1163/15736121-12341342

Kpobi, L., & Swartz, L. (2018a). 'That is how the real mad people behave': Beliefs about and treatment of mental disorders by traditional medicine-men in. *International Journal of Social Psychiatry, 64*(4), 309–316. https://doi.org/10.1177/0020764018763705

Kpobi, L., & Swartz, L. (2018b). 'The threads in his mind have torn': Conceptualization and treatment of mental disorders by neo — Prophetic Christian healers in Accra, Ghana. *International Journal of Mental Health Systems, 1–12*. https://doi.org/10.1186/s13033-018-0222-2

Kpobi, L. N. A., Swartz, L. & Omenyo, C. N. (2019). 'Traditional herbalists' methods of treating mental disorders in Ghana', *Transcultural Psychiatry, 56*(1), 250–266.

Kushitor, M. K., Peterson, M. B., Asante, P. Y., Dodoo, N. D., Boatemaa, S., Awuah, R. B., et al. (2018). Community and individual sense of trust and psychological distress among the urban poor in Accra. *Ghana. PLoS ONE, 13*(9), 1–13. https://doi.org/10.1371/journal.pone.0202818

Lund, C., Breen, A., Flisher, A. J., Kakuma, R., Corrigall, J., Joska, J. A., et al. (2010). Poverty and common mental disorders in low and middle income countries: A systematic review. *Social Science & Medicine, 71*(3), 517–528.

Lazarus, R. S., & Folkman, S. (1984). Stress, appraisal, and coping. New York City: Springer publishing company.

Ohrnberger, J., Fichera, E., & Sutton, M. (2017). The relationship between physical and mental health: A mediation analysis. *Social Science & Medicine, 195*, 42–49.

Opare-Henaku, A., & Utsey, S. O. (2017). Culturally prescribed beliefs about mental illness among the Akan of Ghana. *Transcultural Psychiatry, 54*(4), 502–522.

Organisation for Economic Co-operation and Development. (2010). *Mental health, disability and work*. Paris.

Pham, M. T., Rajić, A., Greig, J. D., Sargeant, J. M., Papadopoulos, A., & McEwen, S. A. (2014). A scoping review of scoping reviews: Advancing the approach and enhancing the consistency. *Research Synthesis Methods, 5*(4), 371–385.

Prince, M., Patel, V., Saxena, S., Maj, M., Maselko, J., Phillips, M. R., & Rahman, A. (2007). No health without mental health. *Lancet, 370*(9590), 859–877. https://doi.org/10.1016/S0140-6736(07)61238-0

Quansah, E., & Karikari, T. K. (2016). Neuroscience-related research in Ghana: A systematic evaluation of direction and capacity. *Metabolic Brain Disease, 31*(1), 11–24.

Read, U. M. (2012). "'I want the one that will heal me completely so it won" t come back again': The limits of antipsychotic medication in rural Ghana. *Transcultural Psychiatry, 49*(3–4), 438–460. https://doi.org/10.1177/1363461512447070

Read, U. M., & Doku, V. C. K. (2012). Mental health research in Ghana: A literature review. *Ghana Medical Journal, 46*(2), 29–38.

Read, U. M., Adiibokah, E., & Nyame, S. (2009). Local suffering and the global discourse of mental health and human rights: An ethnographic study of responses to mental illness in rural. *Globalization and Health, 5*(13), 1–16. https://doi.org/10.1186/1744-8603-5-13

Reilly, J. C., & Houghton, C. (2019). The experiences and perceptions of care in acute settings for patients living with dementia: A qualitative evidence synthesis. *International Journal of Nursing Studies, 96*, 82–90.

Roberts, M., Mogan, C., & Asare, J. B. (2014). An overview of Ghana's mental health system: results from an assessment using the World Health Organization's Assessment Instrument for Mental Health Systems (WHO-AIMS). *International Journal of Mental Health Systems, 8*(16), 1–13.

Salifu Yendork, J., Kpobi, L., & Sarfo, E. A. (2016). "It's only 'madness' that I know": analysis of how mental illness is conceptualised by congregants of selected Charismatic churches in Ghana. *Mental Health, Religion & Culture, 19*(9), 984–999.

Seguin, M., Lewis, R., Razmadze, M., Amirejibi, T., & Roberts, B. (2017). Coping strategies of internally displaced women in Georgia: A qualitative study. *Social Science & Medicine, 194*, 34–41. https://doi.org/10.1016/j.socscimed.2017.10.014

Skinner, E. A., Edge, K., Altman, J., & Sherwood, H. (2003). Searching for the structure of coping: A review and critique of category systems for classifying ways of coping. *Psychological Bulletin, 129*(2), 216–269.

Tawiah, P. E., Adongo, P. B., & Aikins, M. (2015). Mental health-related stigma and discrimination in Ghana: Experience of patients and their caregivers. *Ghana Medical Journal, 49*(1), 30–36.

United Nations. (2015). *General assembly seventieth session* (Vol. 16301).

Walker, G. H., & Osei, A. (2017). Mental health law in Ghana. *BJPsych International, 14*(2), 2011–2012.

World Health Organization. (2003). *Investing in mental health*. Geneva.
World Health Organization. (2008). *mhGAP: Mental health gap action programme: Scaling up care for mental, neurological and substance use disorders*. Geneva.
World Health Organization. (2013). *Mental health action plan: 2013–2020*. Geneva.
World Health Organization. (2017). *Depression and other common mental disorders: Global health estimates*. Global Health Estimates. https://doi.org/CC BY-NC-SA 3.0 IGO

Part II
Maternal Health

Chapter 5
Ghana's Community-Based Health Planning and Services and Women's Decision to Utilize Health Facility-Based Deliveries

Joseph Asumah Braimah, Yujiro Sano, Roger Antabe, and Isaac Luginaah

Introduction

Recognized as a key health challenge of global scale, the Sustainable Development Goal (SDG 3) on health targets reducing maternal mortality to less than 70 maternal deaths per 100,000 births by 2030 (United Nations, 2016). Globally, an estimated 295,000 maternal deaths occurred in 2017, which translated into a maternal mortality ratio (MMR) of 211 per 100,000 live births (WHO, UNICEF, UNFPA, World Bank Group,, & the United Nations Population Division, 2019). Sub-Saharan Africa (SSA) alone accounted for nearly 66% of these deaths with an MMR of 542, which renders it the region with the highest lifetime maternal mortality risk (1 in 37) (WHO et al., 2019). Albeit low, relative to other countries in SSA, Ghana's estimated MMR of 308 underscores the urgent need for improved maternal health care to prevent deaths and meet the SDGs target by 2030. Health facility-based delivery has been identified as vital for reducing delivery complications and preventing maternal deaths (Ghana Health Service, 2017; WHO et al., 2019). In this context, improving health facility-based deliveries is key to reducing maternal mortality in Ghana.

J. A. Braimah (✉) · R. Antabe
Department of Health & Society, University of Toronto Scarborough, Toronto, ON, Canada
e-mail: J.braimah@utoronto.ca; roger.antabe@utoronto.ca

Y. Sano
Department of Sociology, Nipissing University, North Bay, ON, Canada
e-mail: yujiros@nipissingu.ca

I. Luginaah
Department of Geography, Western University, London, ON, Canada
e-mail: iluginaa@uwo.ca

J. A. Braimah et al. (eds.), *Health Geography in Sub-Saharan Africa*, Global Perspectives on Health Geography, https://doi.org/10.1007/978-3-031-37565-1_5

In SSA, women's autonomy has been linked to their assertiveness to sexual and reproductive health, including the utilization of health facility-based deliveries (Dickson et al., 2021; Sano et al., 2018a; Speizer et al., 2014; Woldemicael, 2010). This is because social-cultural norms tend to hinder women from participating in household decisions that affect their lives including the decision to deliver in health facilities (Ganle et al., 2015; Kifle et al., 2018). Moreover, utilizing health facility-based delivery comes at a financial cost and often demands the support and approval of other household members, especially their male partners who tend to control household economic resources that are necessary to finance facility-based deliveries (Kyei-Nimakoh et al., 2017; Zuñiga et al., 2021).

Community-based health initiatives have been identified to be instrumental to promoting women's health, including their health decision-making autonomy (Zuñiga et al., 2021). For instance, Manandhar et al. (2004) observed improved birth outcomes among rural poor in Nepal through a community-based participatory intervention. Also, community health workers were found to help reduce neonatal deaths in rural Bangladesh (Baqui et al., 2009). In unpacking this phenomenon, researchers suggest community-based health programs increase both women and their partners' knowledge on women's reproductive health and rights including the importance of skilled birth delivery (Braimah, 2017; Braimah et al., 2019; Zuñiga et al., 2021). Furthermore, community health initiatives help eliminate geographic barriers to accessing care which have been cited as one of the three delays in accessing care (Adongo et al., 2013; Braimah et al., 2019). The role of community-based health initiatives in facilitating the transportation of patients including pregnant women to utilize health facilities have also been reported in Ghana (Woods et al., 2019). Drawing from these studies, Ghana's Community-based Health Planning and Services (CHPS) strategy may be helpful to enhancing women's ability to independently decide to seek health facility-based deliveries in the country.

The CHPS program is one of the many primary health care initiatives in Ghana to promote universal health coverage. CHPS was adopted in 1999 following the successful pilot of a Community Health and Family Planning project in Navrongo to eliminate geographic and financial barriers to health service use, especially in rural and remote parts of the country (Ministry of Health, 2016). The CHPS initiative focuses on delivering basic health care services to deprived communities through community resource mobilization and engagement by trained community health officers (CHOs) and community health volunteers (CHVs) in a CHPS zone (Adongo et al., 2013; Nyonator et al., 2005). A CHPS zone is a demarcated geographical area, "made up of a town, part of a town or a group of villages/settlements mapped for ease of planning of itinerant services and assignment of CHOs and CHVs" (Ministry of Health, 2016; 22). As a policy directive, it is specified that CHOs educate and empower clients to exercise their rights and motivate pregnant women to utilize skilled delivery (Ministry of Health, 2016). They are also to refer all delivery cases to a higher level of care (Ministry of Health, 2016). Midwifery services may, based on need, be included in the CHPS package. Considering its positioning, the CHPS program can be an important means for enhancing women's health facility-based delivery autonomy.

Women's autonomy to decide to utilize health facility-based deliveries is also dependent on a range of locational, demographic, and economic factors (Bohren et al., 2014; Kifle et al., 2018). Specifically, it has been suggested that women in rural areas tend to face more barriers to deciding independently to deliver in health facilities than their counterparts in urban areas (Dickson et al., 2021). In Busia district in Uganda, having high income status was found to enhance the decision-making ability to utilize skilled delivery services (Anyait et al., 2012). Others include marital status, religious affiliations, and age (Kifle et al., 2018; Kyei-Nimakoh et al., 2017).

This study contributes to the literature by examining the association between residence in functional CHPS zones and women's ability to decide independently to utilize health-facility based delivery services, and whether this varies between rural and urban areas in the Upper West Region (UWR) of Ghana. As of 2017 more than 20% of women lacked access to skilled-delivery services (Ghana Statistical Service, Ghana Health Service, & ICF, 2018). At the same time, only 27% of women were able to make independent decisions concerning their health (Ghana Statistical Service, Ghana Health Service, & ICF International, 2015). The UWR of Ghana is particularly characterized by patriarchy, which has the potential to undermine women's autonomy including the autonomy to decide to seek health facility-based delivery. Specifically, we seek to answer these two questions: (1) What is the relationship between residence in CHPS zones and women's ability to make independent decisions to utilize facility-based delivery? and (2) Do women's ability to independently decide to patronize facility-based deliveries differ between rural and urban areas?

Study Setting

The UWR is located in north-western Ghana and has a population of about 904,695 (Ghana Statistical Service, 2022). The region is largely rural (73.6%) with an agrarian economy, which accounts for 73% of household incomes (Ghana Statistical Service, 2018a, 2022). This among others explains why the region recorded the highest (70.9%) poverty (below the poverty line of = GH¢1314 per adult per year) incidence in Ghana (Ghana Statistical Service, 2018b). Illiteracy rates are high in the region with the rural (70%) areas being worse-off compared to the urban (34.9%) areas. The region is customarily patriarchal and dominated by the extended family system which has ramifications on women's control over household resources and participation in decision-making (Ghana Statistical Service, 2013).

A substantial proportion (17%) of women still lack access to antenatal care despite efforts to ensure a 100% coverage (Ghana Health Service, 2018). Additionally, despite a considerable increase in the proportion of women with access to supervised delivery from 46.5% in 2010 to 68.7% in 2017, it still falls short of the WHO recommendation of achieving 100% health facility-based deliveries (Ghana Health Service, 2018). The institutional maternal mortality ratio in the region is 116.8 deaths per 100,000 live births (Ghana Health Service, 2018). Given these grim health care access statistics, the introduction of CHPS was touted as a positive step to address the health care coverage deficit in the region and Ghana at large (Ministry of Health, 2016; Nyonator et al., 2005). As of 2017, there were 256 CHPS compounds in the region providing various curative and health promotional services (Ghana Health Service, 2018).

Methods

Sampling and Data Collection

Data for this study is cross-sectional and were collected from 440 women in the UWR of Ghana. First, the geo-political districts were clustered into rich and poor using the Ghana Living Standards Survey (GLSS) 6 poverty index. A random sampling technique was then employed to select our study districts—three from the rich districts cluster and four from the poor districts cluster to ensure variability in the data. The number of participants from each district was selected proportional to its population as reported by the 2010 Population and Housing Census of Ghana (Ghana Statistical Service, 2012). Each study district was further clustered into functional and non-functional CHPS zones with 70% and 30% of respondents drawn from them, respectively. A functional CHPS zone is one with an assigned community health officer who provides a defined package of services to the catchment population. A systematic random sampling technique was then employed where every fifth household starting from one end of the sampled community was selected. In each of the households, the woman with the most recent birth was interviewed and where there was none, we skipped. This approach was used to control for possible participant selection biases.

Ethical approval for the study was obtained from the Western University Non-Medical Research Ethics Board. The purpose of the study was clearly explained to participants and their consent orally sought before interviewing. Given the high illiteracy rate in the region, especially among women, oral consent emerged as the most feasible option. However, a spreadsheet with survey ID numbers was prepared for participants to sign/thumbprint as a way of consent to reduce paper waste. Questions for this study were appropriately adapted from the GLSS, a nationwide household survey conducted periodically to generate information on the living conditions of Ghanaians (Ghana Statistical Service, 2014). The modification was done for a number of reasons. First, it has a pool of questions that elicit detail information on households and individuals relevant to this study. Also, while the survey measures health indicators, it does not include questions on decisions on place of delivery, which this study focuses on. Also, the health component focuses mainly on the 2 weeks preceding the interview. Given that the total number of females in the region was 360,928, a minimum of 400 participants were required for statistically significant (95% confidence level) findings (Ghana Statistical Service, 2013; Yamane, 1973).

Measures

For the dependent variable, we asked the women whether they independently decided to deliver in a health facility or not (0 = no; 1 = yes). Senarath and Gunawardena (2009) and Anyait, Mukanga, Oundo, and Nuwaha (Anyait et al., 2012) used a similar question to examine the role of women's autonomy in health care decision-making in South East Asia and Uganda, respectively. The focal independent variable is residence in a functional CHPS zone (0 = yes; 1 = no). Based on previous research (Atuoye et al., 2015; Doku et al., 2012; Speizer et al., 2014), we further included some control variables to account for possible confounding impacts. Our control variables include place of residence, distance to the nearest health facility, household wealth status, level of education, age, marital status, and religious affiliation. A composite index technique was used to construct wealth quintiles using household assets.

Analysis

We employed descriptive and regression analysis. For regression analysis, we employed the logistic regression analysis due to the dichotomous nature of the dependent variable. Models were built sequentially. Model 1 estimates the bivariate relationship between women's decision on place of birth and functional CHPS zone residence, while control variables were further added in Model 2. We also introduced the interaction term between functional CHPS zone residence and urban-rural residence in Model 3 to further understand whether the relationship between women's decision to deliver in a health facility and residence in a CHPS zone differs between rural and urban areas. For meaningful interpretation, the results were reported in odds ratios whereby odds ratios greater than 1 indicate that women were more likely to make an independent decision to deliver in a health facility, while those less than 1 indicate lower odds of doing so.

Results

Among the women that delivered in a health facility, the majority (65%) solely decided to deliver in a health facility (Table 5.1). Three-quarters of them lived in functional CHPS zones with most of them being married (94%). About two-thirds (67%) of the participants lived in rural areas with more than half (54%) of them not having any formal education.

Table 5.2 presents the results from our regression analysis. In Model 1, we found that women in non-functional CHPS zones were less likely to make independent decisions to deliver at a health facility than those in CHPS zones (OR, 0.580; $P < 0.001$). Even after controlling for theoretically relevant variables in Model 2,

Table 5.1 Descriptive analysis of the dependent and independent variables (N = 440)

	Percentage
Made decision to deliver in health facility	
No	35
Yes	65
Residence in functional CHPS zone	
No	25
Yes	75
Urban-rural residence	
Urban	33
Rural	67
Distance to health facility	
Less than 1 km	39
From 1 km to 5 km	24
More than 5 km	37
Household wealth quintiles	
Richest	20
Richer	17
Middle	23
Poorer	20
Poorest	20
Level of education	
Secondary/higher education	23
Primary education	23
No education	54
Age of respondents[a]	28
Marital status	
Currently married	94
Never married	2
Formerly married	4
Religious affiliations	
Christian	57
Muslim	35
Traditionalist	7
No religion	1

[a]Mean reported for age of respondents

living in non-functional CHPS zone was negatively associated with making independent decision to deliver at a health facility (OR = 0.560; $P < 0.001$). These variables include urban-rural residence, distance to the nearest health facility, wealth, education, age, marital status, and religious affiliation. The interaction term introduced in Model 3 further revealed that rural women in non-functional CHPS zones were less likely to make an independent decision to deliver in a health facility when

Table 5.2 Logistic regression analysis predicting 'Women's decision-making autonomy for facility-based delivery' in the Upper West Region of Ghana

	Model 1	Model 2	Model 3
Residence in functional CHPS zone			
Yes	1.00	1.00	1.00
No	0.580 (0.09)***	0.560 (0.09)***	0.95 (0.21)
Urban-rural residence			
Urban		1.00	1.00
Rural		0.76 (0.14)	1.03 (0.21)
Distance to health facility			
Less than 1 km		1.00	1.00
1 to 5 km		0.66 (0.13)*	0.75 (0.15)
More than 5 km		0.59 (0.10)**	0.550 (0.09)***
Household wealth quintiles			
Richest		1.00	1.00
Richer		1.19 (0.26)	1.17 (0.25)
Middle		1.32 (0.31)	1.34 (0.30)
Poorer		1.11 (0.27)	1.08 (0.26)
Poorest		1.17 (0.30)	1.14 (0.29)
Level of education			
Secondary/higher education		1.00	1.00
Primary education		0.80 (0.16)	0.82 (0.16)
No education		0.66 (0.13)*	0.67 (0.13)*
Age of respondents		1.00 (0.01)	1.01 (0.01)
Marital status			
Currently married		1.00	1.00
Never married		1.78 (0.98)	1.59 (0.92)
Formerly married		1.24 (0.44)	1.40 (0.51)
Religion affiliations			
Christian		1.00	1.00
Muslim		0.520 (0.08)***	0.510 (0.08)***
Traditionalist		0.31 (0.12)**	0.36 (0.14)*
No religion		0.46 (0.34)	0.43 (0.32)
Interaction term			
Non-CHPS*rural			0.26 (0.11)**
Log likelihood	−278.67	−243.80	−237.54

$^{*}P < 0.05$, $^{**}P < 0.01$, $^{***}P < 0.001$

compared with their counterparts in non-functional CHPS in urban areas (OR = 0.26; $P < 0.01$).

Some controls were also found to be associated with women's independent decision to deliver in a health facility in the region. For example, compared to women living less than 1 km from a health facility, those living more than 5 km away from a health facility (OR = 0.59; $P < 0.01$) were less likely to make an independent decision to deliver at a health facility. Also, uneducated women were less likely to make an independent decision to deliver at a health facility when compared to those with secondary education or higher (OR = 0.66; $P < 0.05$). Traditionalist (OR = 0.31; $P < 0.01$) and Muslim women (OR = 0.520; $P < 0.001$) were also less likely than their Christian counterparts to report making independent decision to deliver at a health facility.

Discussion and Conclusion

In this study, we examined the differences in facility-based delivery decision-making autonomy between women residing in functional CHPS zones and those outside in the UWR of Ghana. We found that women resident in functional CHPS zones have higher decision-making ability delivering at a health facility than those residing outside. This finding is consistent with previous studies that establish relationship between access to community health promotion programs and women's health decision-making autonomy (Manandhar et al., 2004; Speizer et al., 2014). As the world strives to achieve SDG 3 targets by 2030, it is crucial to take stock of the performances of various community-based primary health care promotion policies, especially those geared at promoting women's reproductive health in the settings of SSA. This is particularly imperative with the persistence of pregnancy and childbirth-related deaths in Ghana, especially in rural and deprived communities (Ghana Health Service, 2018; Ghana Statistical Service et al., 2018a, b). This finding underscores the urgent need to intensify the health promotion activities of community-based facilities given their role in empowering women to make an independent decision on delivering at a health facility. Indeed, earlier studies have also elaborated on how decision-making power within the household setting is essential to women's assertiveness to sexual and reproductive health (Ngom et al., 2003; Sano et al., 2018a).

There are several mechanisms through which residence in functional CHPS zones can lead to facility-based delivery autonomy. For instance, the CHPS policy framework mandates CHOs and CHVs to provide antenatal care services alongside their preventive and curative duties (Ministry of Health, 2014; Shamsu-Deen, 2015). Health education and sensitization provided through antenatal care can empower women to decide to utilize facility-based deliveries (Shamsu-Deen, 2015). Thus, CHOs and CHVs through their house-to-house visits and community engagements have the potential to change negative attitudes and eliminate sociocultural barriers that hinder women from exercising their health rights including the autonomy to decide on place of delivery. This assertion is affirmed by Adongo et al. (2013) who observed that males are more involved in family planning services in communities with CHPS facilities than those without CHPS in southern Ghana. Similarly, in Navrongo in the Upper East Region of Ghana, the CHPS initiative was found to have eliminated the traditional household decision-making structure, also referred to as the gate-keeping system where men dictate women's use of health services

(Ngom et al., 2003). Thus, through their involvement and management of their health care system—the founding element of the CHPS program—along with its educational and advocacy programs, men in functional CHPS zones may become more acquainted, sensitive, and involved in women's health and granting them the autonomy to make health decisions including where to deliver their babies.

Social and human capital engendered by the program may also be important for improving women's reproductive autonomy and health service use in the region. Social capital is associated with improved health and health service use, especially in resource constrained settings. CHOs and CHVs as a mandate also engage in community mobilization programs such as mother-to-mother support groups and community transport systems to empower and promote health service use (Ministry of Health, 2016). Similar to our finding, Woods et al. (2018) found in the UWR of Ghana that CHPS contributed significantly to human, social, and economic capital in communities which considerably improved health and health service use including skilled delivery.

Our interaction term shows that the odds of independently deciding to deliver in a health facility is lower among rural women in non-functional CHPS zones compared with urban women in non-functional CHPS zones. This finding is consistent with previous studies that highlight low autonomy among rural women in low- and middle-income countries (Anyait et al., 2012; Bohren et al., 2014). Although the CHPS program was implemented to bridge the health service access gap between rural and urban areas, many of these rural areas still lack essential health care services and capacity building opportunities. Rural women continue to experience worse forms of marginalization, poverty, and illiteracy, which have implications for their sexual and reproductive health decision-making. While CHOs provide outreach services to rural communities, the geographies of these rural areas coupled with resource constraints impede the provision of appropriate health care services. Traditional and patriarchal household practices that perpetuate low autonomy among women are likely to be stronger in rural areas than urban areas possibly explaining our findings (Atuoye et al., 2015; Braimah et al., 2019).

While this analysis deepens our understanding of the link between residence in CHPS zone and women's independent decision-making ability to utilize facility-based delivery, there are some limitations worth noting. First, the cross-sectional nature of the study limits the ability to establish causal relationship between residence in functional CHPS zone and women's decision-making autonomy regarding place of birth, hence limiting our findings to only statistical associations. Additionally, gender issues including women's sexual and reproductive rights and health are sensitive, and their responses may suffer from social desirability bias.

Our findings demonstrate that the CHPS program may be helpful in promoting women's reproductive autonomy in the UWR of Ghana. Indeed, understanding women's decision-making autonomy for facility-based delivery in the context of the CHPS initiative is important for accelerating Ghana's efforts toward achieving universal health coverage and the on-going sustainable development goals. Also, the study joins previous studies to reiterate existing health care access disparities between rural and urban areas in Ghana. Beyond that, the findings reinforce the understanding that the decision to deliver in a health facility is multilayered and transcends the individual and their personal characteristics to include broader structural factors. Drawing from these findings, we make the following recommendations. Firstly, it will be beneficial to incorporate programs that bridge gender disparities and enhance women's health decision-making autonomy within the CHPS model in Ghana. Our analysis also pointed out that rural women outside of functional CHPS zone may struggle establishing their autonomy for facility-based delivery. Policymakers should be aware that rural women may not be benefitting from formalized systems that guarantee greater gender balance in decision-making including community-based education programs and media campaigns that can be expanded with the effort of the CHPS initiative. Future studies may also draw on longitudinal and qualitative research to deeply explore the relationship between women's decision on place of birth and residence in functional CHPS zones and how their reproductive and sexual rights can be improved.

Acknowledgments We are grateful to the Western Graduate Research Scholarship and Dr. Isaac Luginaah's (Canada Research Chair in Health Geography) for supporting this study. We are also thankful to the research participants for agreeing to take part in this study. Also, to the research assistants we say thank you for your help in collecting the data.

References

Adongo, P. B., Tapsoba, P., Phillips, J. F., Tabong, P. T.-N., Stone, A., Kuffour, E., et al. (2013). The role of community-based health planning and services strategy in involving males in the provision of family planning services: A qualitative study in southern Ghana. *Reproductive Health, 10*(36), 1–15.

Anyait, A., Mukanga, D., Oundo, G. B., & Nuwaha, F. (2012). Predictors for health facility delivery in Busia district of Uganda: A cross sectional study. *BMC Pregnancy and Childbirth, 12*(1), 1.

Atuoye, K. N., Dixon, J., Rishworth, A., Galaa, S. Z., Boamah, S. A., & Luginaah, I. (2015). Can she make it? Transportation barriers to accessing maternal and child health care services in rural Ghana. *BMC Health Services Research, 15*(1), 333.

Baqui, A. H., Arifeen, S. E., Williams, E. K., Ahmed, S., Mannan, I., Rahman, S. M., et al. (2009). Effectiveness of home-based management of newborn infections by community health workers in rural Bangladesh. *The Pediatric Infectious Disease Journal, 28*(4), 304.

Bohren, M. A., Hunter, E. C., Munthe-Kaas, H. M., Souza, J. P., Vogel, J. P., & Gülmezoglu, A. M. (2014). Facilitators and barriers to facility-based delivery in low- and middle-income countries: A qualitative evidence synthesis. *Reproductive Health, 11*(71), 1–17. https://doi.org/10.1186/1742-4755-11-71

Braimah, J. A. (2017). *Community-based health planning and services and women's access to health care in the upper west region of Ghana* (Masters thesis). Western University.

Braimah, J. A., Sano, Y., Atuoye, K. N., & Luginaah, I. (2019). Access to primary health care among women: The role of Ghana's community-based health planning and services policy. *Primary Health Care Research & Development, 20*(e82), 1–7.

Dickson, K. S., Adde, K. S., & Ameyaw, E. K. (2021). Women empowerment and skilled birth attendance in sub-Saharan Africa: A multi-country analysis. *PLoS One, 16*(7).

Doku, D., Neupane, S., & Doku, P. N. (2012). Factors associated with reproductive health care utilization among Ghanaian women. *BMC International Health and Human Rights, 12*(1), 1.

Ganle, J. K., Obeng, B., Segbefia, A. Y., Mwinyuri, V., Yeboah, J. Y., & Baatiema, L. (2015). How intra-familial decision-making affects women's access to, and use of maternal healthcare services in Ghana: A qualitative study. *BMC Pregnancy & Childbirth, 15*(1), 1.

Ghana Health Service. (2017). *Ghana health service 2016 annual report*. Accra. https://doi.org/10.1136/bjo.2010.193169.

Ghana Health Service. (2018). *The health sector in Ghana: Facts and figures*. Accra.

Ghana Statistical Service. (2012). *2010 population and housing census. Ghana statistical service*. Accra, Ghana. https://doi.org/10.1371/journal.pone.0104053.

Ghana Statistical Service. (2013). *2010 population & housing census: Regional analytical report, upper west region*. Accra, Ghana.

Ghana Statistical Service. (2014). *Ghana living standards survey round 6 (GLSS 6): Main report*. Accra, Ghana. https://doi.org/10.1016/j.infsof.2008.09.005.

Ghana Statistical Service. (2018a). *Ghana living standards survey (Round 7)*. Accra, Ghana.

Ghana Statistical Service. (2018b). Ghana living standards survey (GLSS7): Poverty trends in Ghana; 2005–2017. *Ghana statistical service*, 1–62.

Ghana Statistical Service. (2022). *Ghana 2021 population and housing census: General report highlights*. Accra.

Ghana Statistical Service, Ghana Health Service, & ICF. (2018). *Ghana maternal health survey 2017*. Accra, Ghana.

Ghana Statistical Service, Ghana Health Service, & ICF International. (2015). *Ghana demographic and health survey.*

Kifle, M. M., Kesete, H. F., Gaim, H. T., Angosom, G. S., & Araya, M. B. (2018). Health facility or home delivery? Factors influencing the choice of delivery place among mothers living in rural communities of Eritrea. *Journal of Health, Population and Nutrition, 37*(22), 1–15. https://doi.org/10.1186/s41043-018-0153-1

Kyei-Nimakoh, M., Carolan-Olah, M., & McCann, T. V. (2017). Access barriers to obstetric care at health facilities in sub-Saharan Africa-a systematic review. *Systematic Reviews, 6*(110), 1–16. https://doi.org/10.1186/s13643-017-0503-x

Manandhar, D. S., Osrin, D., Shrestha, B. P., Mesko, N., Morrison, J., Tumbahangphe, K. M., et al. (2004). Effect of a participatory intervention with women's groups on birth outcomes in Nepal: Cluster-randomised controlled trial. *The Lancet, 364*(9438), 970–979.

Ministry of Health. (2014). *National community health planning and services*. Accra, Ghana.

Ministry of Health. (2016). *National community-based health planning and service (CHPS) Policy*. Accra, Ghana.

Ngom, P., Debpuur, C., Akweongo, P., Adongo, P., & Binka, F. N. (2003). Gate-keeping and women's health seeking behaviour in Navrongo, northern Ghana. *African Journal of Reproductive Health, 7*(1), 17–26.

Nyonator, F. K., Awoonor-Williams, J. K., Phillips, J. F., Jones, T. C., & Miller, R. A. (2005). The Ghana community-based health planning and services initiative for scaling up service delivery innovation. *Health Policy and Planning, 20*(1), 25–34.

Sano, Y., Antabe, R., Atuoye, K. N., Braimah, J. A., Galaa, S. Z., & Luginaah, I. (2018a). Married women's autonomy and post-delivery modern contraceptive use in the Democratic Republic of Congo. *BMC Women's Health, 18*(1), 49.

Sano, Y., Sedziafa, A. P., Vercillo, S., Antabe, R., & Luginaah, I. (2018b). Women's household decision-making autonomy and safer sex negotiation in Nigeria: An analysis of the Nigeria demographic and health survey. *AIDS Care, 30*(2), 240–245.

Senarath, U., & Gunawardena, N. S. (2009). Women's autonomy in decision making for health care in South Asia. *Asia-Pacific Journal of Public Health, 21*(2), 137–143.

Shamsu-Deen, Z. (2015). Bringing health care to the doorstep of the rural folk in Ghana: Assessing the potentials and challenges of CHIPS. *Indian Journal of Pharmaceutical Science & Research, 5*(1), 1–9.

Speizer, I. S., Story, W. T., & Singh, K. (2014). Factors associated with institutional delivery in Ghana: The role of decision-making autonomy and community norms. *BMC Pregnancy and Childbirth, 14*(1), 1.

United Nations. (2016). *The sustainable development goals report.*

WHO, UNICEF, UNFPA, World Bank Group, & the United Nations Population Division. (2019). *Trends in maternal mortality 2000 to 2017*. Geneva.

Woldemicael, G. (2010). Do women with higher autonomy seek more maternal health care? Evidence from Eritrea and Ethiopia. *Health Care for Women International, 31*(7), 599–620.

Woods, H., Haruna, U., Konkor, I., & Luginaah, I. (2018). The influence of the community-based health planning and services (CHPS) program on community health sustainability in the upper west region of Ghana. *The International Journal of Health Planning and Management*, 1–15.

Woods, H., Haruna, U., Konkor, I., & Luginaah, I. (2019). The influence of the community-based health planning and services (CHPS) program on community health sustainability in the upper west region of Ghana. *The International Journal of Health Planning and Management, 34*(1), e802–e816. https://doi.org/10.1002/hpm.2694

Yamane, T. (1973). *Statistics: An introductory analysis* (3rd ed.). Harper and Row.

Zuñiga, J. A., García, A., Kyololo, O. M., Hamilton-Solum, P., Kabimba, A., Milimo, B., et al. (2021). Increasing utilisation of skilled attendants at birth in sub-Saharan Africa: A systematic review of interventions. *International Journal of Nursing Studies, 120*(2021), 103977. https://doi.org/10.1016/j.ijnurstu.2021.103977

Chapter 6
Postpartum Married Women's Mass Media Exposure to Family Planning Messages in the Democratic Republic of Congo: The Role of Women's Household Decision-Making Autonomy

Florence Wullo Anfaara, Roger Antabe, and Yujiro Sano

Introduction

Sub-Saharan Africa (SSA) faces the highest maternal death rate in the world. In 2017, the global average for maternal deaths was estimated at 211 deaths per every 100,000 births. In SSA, however, maternal deaths were 542 per 100,000 births—constituting 331 more maternal deaths on average than in any other part of the world (WHO, UNICEF, UNFPA, World BANK, & UN, 2019). The situation is daunting, especially for SSA countries that have experienced conflict and political instability for many years. Although effective postpartum contraceptive use and family planning are associated with positive reproductive health outcomes for women, including the reduction in maternal and child mortality rates (Cleland et al., 2012), conflict-affected countries such as the Democratic Republic of Congo (DRC) are still struggling to meet the demands for modern contraceptive use (Corker, 2010; Tran et al., 2020). In 2014, for instance, the DRC Demographic and Health Survey reported about 28% of unmet needs for adult contraception use (CDHS, 2014). Also, the DRC's maternal mortality deaths of 693 per 100,000 in 2014 is three times

F. W. Anfaara (✉)
Department of Gender, Sexuality, and Women's Studies, Western University, London, ON, Canada
e-mail: fanfaara@uwo.ca

R. Antabe
Department of Health and Society, University of Toronto Scarborough, Toronto, ON, Canada
e-mail: roger.antabe@utoronto.ca

Y. Sano
Department of Sociology, Nipissing University, North Bay, ON, Canada
e-mail: yujirosano@nipissingu.ca

J. A. Braimah et al. (eds.), *Health Geography in Sub-Saharan Africa*, Global Perspectives on Health Geography, https://doi.org/10.1007/978-3-031-37565-1_6

higher than the global average (Kwete et al., 2018; WHO et al., 2019). In addition to its high maternal death rate, the DRC records other maternal and child health issues, such as neonatal and infant deaths, unintended pregnancies, short child spacing, and poor nutrition (Lawry et al., 2019). Not only are maternal and reproductive health issues a problem in the DRC, but they are also a part of a broader global public health concern. Consequently, as part of a global effort to reduce maternal and child deaths, the UN Sustainable Development Goal (SDG) 3.7 has been dedicated to addressing this issue. The aim of SDG 3.7 is to reduce maternal mortality rate by 44% in 2030 through the promotion of modern contraceptive use prioritizing its availability and access to women of reproductive age but especially during postpartum periods (Ahmed et al., 2012; Osotimehin, 2015; Starbird et al., 2016).

With decades of conflict and the recent Ebola outbreak, the health infrastructure of the DRC is depleted, with several locations lacking properly trained health personnel providing modern contraception information to women of reproductive ages and postpartum women (Corker, 2010). One of the steps to addressing this high maternal death rate was the adoption and integration of the World Health Organization's postpartum family planning (PPFP) programming into its local and national health plans under the 6-year national family planning strategic plan (2014–2020). When rolled out, the DRC Ministry of Health expects the PPFP programming to increase access and use of modern contraceptive to increase to 13% by 2020 (Mikatho et al., 2021; DRC Ministry of Health, 2014). Notwithstanding the challenges, the DRC recorded a slight increase in contraceptive use by 8% towards the end of 2014. Nonetheless, access and availability of modern contraceptive use remains challenging among postpartum women in the DRC and rural areas, making it difficult to achieve the country's PPFP target.

As a result, mass media (e.g., radio, television, and print media) has taken up the role of reaching many people with information on modern contraceptive use in the DRC. Mass media messaging has been reported to positively influence health behaviors, including the uptake of modern contraceptive use. For example, a study conducted in Bangladesh by Islam and Hasan (2000) found that radio and television family planning messages influenced the uptake and use of modern contraception. Similarly, Bankole and Westoff (1996) found that exposure to mass media predicted the uptake and use of modern contraception and family planning. Within the study context, a Kinshasa-based non-governmental organization designed and implemented family planning messages via radio and television, which witnessed an increase in contraceptive use from 10.8% to 25.7% in 2015 (Development Media International, 2015). The role of mass media exposure and the uptake of modern contraceptive use in DRC cannot be overemphasized. For instance, it was reported that through a nationwide media campaigns on family planning and modern contraceptive use, women in remote and unmotorable areas who would have hitherto not had access to maternal health information were reached (Hernandez et al., 2018; Ministére de la santé publique, 2014; Mukaba et al., 2015). In addition, the government and its development partners have emphasized the importance of mass media campaigns in improving women's knowledge and uptake of modern contraception in the country (Kayembe et al., 2006).

Despite its important role in promoting modern contraceptive use, research shows that mass media information about family planning is not equally accessible to most women due to demographic, socioeconomic, and sociocultural factors. For instance, Konkor et al. (2019) found in Nigeria that women experienced differential exposures to modern contraception messages based on rural-urban residence, level of education, and employment status. These factors tend to influence the frequency with which women are exposed to mass media messaging on modern contraceptives. Similarly, Ajaero et al. (2016) found that educated and richer women were more likely to use modern contraception than their lower-status counterparts. Similarly, in Bangladesh, it was found that women in urban areas who were more exposed to mass media messages on family planning reported higher usage of modern contraception relative to their rural counterparts (Jahan et al., 2017).

While these factors are important, the role of gender dynamics on married women's household decision-making autonomy (hereafter called women's autonomy) and exposure to mass media family planning messaging is largely missing in the literature. Although there is no standardized definition for women's autonomy, it often signifies married women's ability to make independent decisions on a range of issues concerning their lives and those of their dependents without recourse to permission from their partners (Mason, 1986). Research shows that married women with higher levels of autonomy tend to have greater control over their health choices and needs, such as access to health information and modern contraceptives (Antabe et al., 2021; Sano et al., 2018b). In explaining this phenomenon, earlier scholars have pointed to the unique gendered power dynamics within the household where power on decision-making and personal choices is not skewed in favor of husbands (Basu, 1992; Dyson & Moore, 1983). However, like with other SSA countries, the DRC continues to exhibit traditional gender norms at the household level that tend to favor men over women. Therefore, women's access to health information, including family planning messages, may be compromised (Konkor et al., 2019). In addition, the desirability for large family sizes in traditional sociocultural contexts, including the DRC, can potentially make it challenging for married women with limited autonomy to consume health information on family planning through mass media (Muanda et al., 2016).

Despite its importance, the relationship between women's autonomy and mass media exposure to family planning messages may vary based on urban-rural locational characteristics (Jahan et al., 2017). In the DRC, social infrastructural resources, including mass media, are unequally available between urban and rural areas. Specifically, the concentration of family planning programs and mass media resources such as television, radio, and newspaper in urban areas makes access to health information through these mass media mediums more easily available to urban residents (Corker, 2010; Kwete et al., 2018). In contrast, the scarcity of mass media resources in rural settings may often limit married women's exposure to and interaction with health information disseminated through mass media (Jacobs et al., 2017). Consequently, household-gender dynamics may be at play in how married women can access health information through mass media at their convenience in rural areas. Furthermore, rural areas tend to conserve traditional culture that endorse large family sizes and also prioritizes men's decisions over women's within the household (Corker, 2010; Muanda et al., 2016). In such contexts, improving women's autonomy may be particularly critical for rural women in tackling sociocultural norms that may limit their reproductive choices, including use of health information through mass media.

To this end, despite growing scholarly and policy interest in health information dissemination through mass media in SSA and the DRC, we know very little about whether household-level gender power dynamics between wives and husbands can impact married women's mass media exposure to family planning messages. To address this void in the literature, we use a nationally representative survey data to examine: (1) the association between married women's autonomy and their mass media exposure to family planning messages in the DRC and (2) the potential moderating role of geography defined by urban-rural residence.

Methods

To understand the role of women's autonomy and mass media exposure to family planning messages, we used the 2013–14 Democratic Republic of Congo Demographic and Health Survey (CDHS). The CDHS is a nationally representative dataset of women aged 15–49, implemented by the Ministry of Monitoring, Planning, and Implementation of the Modern Revolution, in collaboration with the Ministry of Public Health. The CDHS utilized a multi-staged sampling framework in which systematic sampling with probability to size was applied to identify Enumeration Areas from which households were chosen. A total of 19,097 women were initially identified at the time of fieldwork, and 18,827 women were successfully interviewed, with a response rate of 99%. We limited our sample to married women who have given birth in the last 3 years for two important reasons. For one, women's autonomy is a concept that captures the power dynamics between wife and husband. For another, the WHO (2006) recommends women practice birth spacing, waiting at least 2-3 years to have another pregnancy after the most recent delivery. Accordingly, our analytical sample includes 8015 married women whose recent birth was within the last 3 years.

Measures

The dependent variable is exposure to mass media family planning 'message.' Specifically, women were asked whether they have heard any messages about family planning through (1) radio, (2) television, and (3) newspaper and magazine (0 = no; 1 = yes). Based on these questions, we created a single binary variable called 'mass media exposure to family planning messages.' If a respondent answered 'yes' to having been exposed to one of the above mentioned media sources, we categorized such responses as 1, and for those who chose 'no' we coded it 0 (0 = no; 1 = yes).

Four relevant questions capture the power dynamics between wife and husband within their household. Specifically, the CDHS asked married women about who usually has the final say in household settings on the following decisions: (1) obtaining their health care, (2) making large purchases for daily needs, (3) visiting family and relatives, and (4) household expenditures (0 = husband alone; 1 = respondent and husband together; 2 = the respondent alone). Using these four questions, we use principal component analysis to construct a summative scale called 'women's autonomy' as our focal independent variable. We found this scale to be statistically robust and reliable ($\alpha = 0.74$). All indicators were loaded on a single latent scale construct with factor loadings, ranging from 0.70 to 0.79. Higher values on this scale indicate that women have higher autonomy while lower values indicate lower autonomy.

Based on previous studies (Kansanga et al., 2018; Konkor et al., 2019), we account for two sets of factors that may impact married women's exposure to family planning on mass media. For one, we control for location and demographic factors such as place of residence (0 = urban; 1 = rural), province of residence (0 = Kinshasa; 1 = Bandundu; 2 = Bas-Congo; 3 = Equateur; 4 = Kasai-Occidental; 5 = Kasai-Oriental; 6 = Katanga; 7 = Maniema; 8 = Nord-Kivu; 9 = Orientale; 10 = Sud-Kivu), age of respondents (measured as a continuous variable), religious affiliation (0 = Christian; 1 = non-Christian), and number of births (0 = less than three; 1 = three to five; 2 = more than five). For another, we include three socioeconomic indicators such as household income (0 = poorest; 1 = poorer; 2 = middle; 3 = richer; 4 = richest), level of education (0 = no education; 1 = primary education; 2 = secondary or higher education), and employment status (0 = not employed; 1 = employed).

Statistical Analysis

There are two separate analyses for this study. First, we employ univariate analysis to describe the characteristics of the analytical sample. Second, due to the binary nature of the dependent variable (Menard, 2002), we rely on logistic regression analysis to understand (1) the association between women's autonomy and mass media family planning message exposure and (2) whether the association between the two variables differs between rural and urban areas. We estimate the bivariate impact of women's autonomy on mass media message exposure in Model 1 and sequentially account for demographic and socioeconomic factors in Models 2 and 3, respectively. We introduce the interaction term between women's autonomy and place of residence in Model 4. Due to the multistage sample design, complex population-based surveys such as the CDHS usually have some degree of dependence among the observations. This can be problematic because standard regression models such as logistics regressions often assume independence among the observations (Raudenbush & Bryk, 2002). Using the '*vce*' option provided by STATA, we account for this potential bias by imposing a 'cluster' variable, which constitutes the unique identification numbers assigned to respondents at the cluster level. Results are shown in odds ratios (ORs). ORs larger than 1 indicate that married women are more likely to be exposed to family planning messages through mass media, while those smaller than 1 imply lower odds of being exposed to family planning messages. We apply sampling weights provided in the data to all the analyses.

Results

Table 6.1 shows findings from univariate analysis. We find that 11% of post-delivery married women have been exposed to family planning messages through radio, television, and print media. The majority of respondents are Christian (96%), employed (76%), and live in rural areas (71%). About one-third also report having less than three births (31%) and one-fifth (19%) do not have formal education. The largest proportion of respondents live in Bandundu (17%), followed by Equateur (14%), Kasai-Oriental (12%), and Katanga (11%). The mean age of respondents is 28.78 years old.

Table 6.2 shows findings from regression analysis. We find in Model 1 that married women's autonomy is positively and significantly associated with married women's exposure to family planning messages at the bivariate level (OR = 1.35, $p < 0.001$), suggesting that women with higher levels of autonomy are more likely to be exposed to family planning messages through mass media than those with lower levels. When we control for location and demographic variables in Model 2, the impact of women's autonomy on their mass media exposure to family planning messages is partly attenuated (OR = 1.16, $p < 0.01$). Further investigation shows place of residence to be responsible for the observed attenuation of the relationship between women's autonomy and mass media exposure to family planning messages. Furthermore, Model 3 suggests that the role of women's autonomy on media exposure to family planning messages is partly attenuated when we further control for socioeconomic factors (OR = 1.16, $p < 0.05$). In Model 4, we introduce the interaction term between women's autonomy and place of residence and find that the impact of women's autonomy on mass media exposure to family planning messages is greater in rural than urban areas (OR = 1.35, $p < 0.01$).

In addition, we find a range of control variables to be significantly associated with married women's mass media exposure to family planning messages (see Model 3 of Table 6.2). For example, rural women are less likely to be exposed to family planning messages than those in urban areas (OR = 0.35, $p < 0.001$). Compared to those in Kinshasa, women in Bandundu (OR = 2.37, $p < 0.001$), Maniema (OR = 2.72, $p < 0.001$), Nord-Kivu (OR = 6.35, $p < 0.001$), and Sud-Kivu (OR = 4.63, $p < 0.001$) were respectively more likely to be exposed to family planning messages. In addition, married women with more than five children are less

Table 6.1 Univariate analysis of the dependent and independent variables

	Percentage
Exposure to family planning message	
No	89
Yes	11
Place of residence	
Urban	30
Rural	70
Women's autonomy[a]	0.04
Age of respondents[a]	28.78
Religious affiliation	
Christian	96
Non-Christian	4
Province of residence	
Kinshasa	7
Bandundu	17
Bas-Congo	4
Equateur	14
Kasai-Occidental	8
Kasai-Oriental	12
Katanga	11
Maniema	3
Nord-Kivu	8
Orientale	8
Sud-Kivu	8
Place of residence	
Urban	29
Rural	71
Number of births	
Less than three	31
Three to five	41
More than five	28
Household income	
Poorest	22
Poorer	23
Middle	20
Richer	18
Richest	17
Level of education	
No education	19
Primary education	44
Secondary/higher education	37
Employment status	
Not employed	24
Employed	76
Total	8015

[a]Mean score is reported for continuous variable

Table 6.2 Logistic regression analyzing 'exposure to family planning message' among post-delivery women in the DRC

	Model 1	Model 2	Model 3	Model 4
	OR (SE)	OR (SE)	OR (SE)	OR (SE)
Women's autonomy	1.35 (0.06)***	1.16 (0.06)**	1.16 (0.05)*	1.02 (0.07)
Place of residence				
Urban		1.00	1.00	1.00
Rural		0.14 (0.02)***	0.35 (0.06)***	0.33 (0.06)***
Women's autonomy*rural				1.35 (0.16)**
Province of residence				
Kinshasa		1.00	1.00	1.00
Bandundu		1.36 (0.27)	2.37 (0.56)***	2.23 (0.53)**
Bas-Congo		0.92 (0.29)	1.06 (0.34)	0.98 (0.31)
Equateur		0.38 (0.09)***	0.77 (0.20)	0.73 (0.19)
Kasai-Occidental		0.57 (0.15)*	1.06 (0.31)	1.01 (0.29)
Kasai-Oriental		0.62 (0.12)*	0.97 (0.21)	0.92 (0.20)
Katanga		0.87 (0.19)*	1.08 (0.25)	1.01 (0.23)
Maniema		1.37 (0.30)	2.72 (0.72)***	2.60 (0.69)***
Nord-Kivu		3.71 (0.78)***	6.35 (1.53)***	5.98 (1.45)***
Orientale		0.82 (0.20)	1.37 (0.36)	1.29 (0.34)
Sud-Kivu		3.69 (1.04)***	4.63 (1.36)***	4.14 (1.19)***
Age of respondents		1.02 (0.01)	1.01 (0.01)	1.01 (0.01)
Religious affiliation				
Christian		1.00	1.00	1.00
Non-Christian		0.78 (0.20)	0.86 (0.23)	0.87 (0.23)
Number of births				
Less than three		1.00	1.00	1.00
Three to five		0.79 (0.13)	0.93 (0.14)	0.94 (0.14)
More than five		0.46 (0.10)**	0.61 (0.14)*	0.61 (0.14)*
Household income				
Poorest			1.00	1.00
Poorer			1.18 (0.32)	1.16 (0.32)
Middle			2.04 (0.53)**	2.06 (0.54)**
Richer			2.87 (0.75)***	2.87 (0.75)***
Richest			4.47 (1.30)***	4.49 (1.31)***
Level of education				
No education			1.00	1.00
Primary education			1.10 (0.26)	1.13 (0.26)
Secondary/higher education			2.56 (0.64)***	2.61 (0.65)***
Employment status				
Not employed			1.00	1.00
Employed			0.99 (0.13)	0.99 (0.13)
Log pseudolikelihood	−2714.22	−2267.04	−2160.02	−2154.00
Wald x2	44.85***	489.02***	538.49***	528.77***
Pseudo r2	0.01	0.18	0.21	0.22

OR odds ratio, *SE* standard error
$^*p < 0.05$, $^{**}p < 0.01$, $^{***}p < 0.001$

likely to be exposed to family planning messages through mass media than those with less than three children (OR = 0.61, $p < 0.05$). We also find two socioeconomic factors to be significantly associated with mass media exposure to family planning messages. Specifically, women's exposure to mass media family planning messages increases as household wealth also increases. In addition, compared to those without any formal education, women with secondary or higher education are more likely to be exposed to family planning messages through mass media (OR = 2.56, $p < 0.001$).

Discussion

Given the DRC's inability to meet the Millennium Development Goal 5 target of reducing maternal deaths by 75% by 2015, the government has re-strategized to meet the SDG 3.7 target by aiming to increase the use of modern contraceptives by 13% at the end of 2020 through its National Strategic Plan (DRC Ministry of Health, 2014; WHO et al., 2019). However, the contextual dynamics of years of civil war and poorly developed social infrastructure have made it difficult to reach especially postpartum women in all parts of the country with messages on modern contraceptive use. In this regard, mass media has become an important medium to reach and encourage postpartum women to use modern contraception in the DRC (Kandala et al., 2014). However, despite its importance, we know very little about how married women's mass media exposure to family planning messages is impacted by their autonomy in their postpartum periods and whether this relationship varies between urban and rural areas. Using the 2013–14 Democratic Republic of Congo Demographic and Health Survey, the current study addressed these voids in the literature.

Our findings show that increased women's autonomy positively impacts exposure to mass media messages about family planning during postpartum periods. This revelation is consistent with earlier research, identifying women's autonomy as a critical determinant of a range of desirable health behaviors such as safer sexual practices, antenatal care utilization, and avoidance of unintended pregnancy (Abada & Tenkorang, 2012; Amoyaw et al., 2015; Woldemicael & Tenkorang, 2010). These studies underscore how autonomous women in highly traditional contexts can become more assertive about their healthcare needs and might have found ways to deal with unbalanced gendered power dynamics within household settings giving them the ability to exercise control over their needs including the use of family planning messages through mass media. This trend may be helpful in the DRC, especially where women's mass media exposure to health information is potentially neglected by the emphasis on prioritization of men's decisions within household settings, persistently subordinating women's independent decision-making about their health and health needs to that of their male partners (Muanda et al., 2016).

We also observe that the role of autonomy on married women's mass media exposure to family planning messages is largely attenuated when we introduced demographic and socioeconomic factors, particularly education, household wealth, and urban-rural residency. These findings imply that having lower socioeconomic status and living in rural communities can serve as a barrier to attaining higher levels of women's autonomy, which can further be detrimental to their mass media exposure to family planning messages. This is consistent with previous research (Boateng et al., 2014), suggesting that women's socioeconomic status plays an important role in their ability to participate in household decision-making processes and ensure that their opinions are represented within household settings. Similarly, Nigatu et al. (2014) identify in Ethiopia that rural women were less likely to be autonomous than their urban counterparts, possibly signifying that some conditions of rural environments may not encourage women to make independent decisions within household settings.

Importantly, we also find that the impact of autonomy on married women's mass media exposure to family planning messages is greater for rural women relative to those in urban areas. This finding may be explained by the nature of rural areas in the DRC, often characterized by the poor deployment of mass media resources in addition to the persistence of traditional culture that enforces men's decisions over that of women within household settings (Corker, 2010; Kwete et al., 2018). Specifically, these contextual dynamics may serve as a platform for women's autonomy to become an indispensable tool to which married women potentially navigate existing patriarchal structures and norms that may hinder their mass media exposure to family planning messages (Alemayehu & Meskele, 2017). Consequently, consistent with our expectation, women's autonomy appears more useful for rural married women to meet parts of their healthcare needs during postpartum periods through mass media exposure relative to their urban counterparts.

While not the focus of our study, we also observe that married women's exposure to family planning is impacted by several demographic and socioeconomic factors. For example, compared to women in Kinshasa, those in other provinces such as Bandudu, Maniema, Nord-Kivu, and Sud-Kivu are more likely to be exposed to mass media messages on family planning. Health information dissemination through mass media in SSA often targets people in remote and inaccessible locations (Sano et al., 2016a, b). Media campaigns on family planning in the DRC organized by NGOs may be serving as a crucial source of reproductive health care information for postpartum women in these provinces (Alberti et al., 2010). We also find that, compared to poorer women, richer women are more likely to be exposed to mass media messages on family planning. This result is consistent with findings by Konkor et al. (2019), that greater household wealth facilitated women's purchases and access to mass media outlets such as radio and television. We also find that educated women are more likely to be exposed to family planning messages through mass media than uneducated ones. This observation may be explained by Sano et al. (2018a, b), arguing that formal education enables people to be more assertive about their health and healthcare needs.

We make some noteworthy limitations to our study. First, due to the cross-sectional nature of the CDHS, our study findings are limited to a statistical association. Therefore, the interpretation of the findings should be done with caution.

Second, postpartum women may be receiving information on modern contraception besides those classified in this study and the CDHS as mass media. For instance, religious and non-governmental organizations in SSA are important health information and promotion sources that can influence women's exposure to modern contraception messages (Anfaara et al., 2020). Additionally, in the patriarchal context of the DRC, questions about autonomy and contraception may be sensitive, and this could potentially lead to bias in the responses. To avoid these limitations, future research needs to assess other potential sources of health information on modern contraceptives and family planning for postpartum women in the DRC. It may also be useful for future studies to examine the effectiveness of these sources on postpartum women's knowledge and uptake of modern contraceptives as the DRC aims to meet SDG 3.7 (Lee et al., 2019). Despite these limitations, our study makes an important contribution to both policy and the literature on how autonomy and a range of demographic and socioeconomic factors shape postpartum married women's mass media exposure to family planning messages.

Based on our findings, we make some important policy suggestions. First, given that married women's autonomy is associated with their exposure to mass media messages on family planning in the DRC, there is an urgent need for policy stakeholders to prioritize improving women's autonomy as part of a larger strategy to meet the SDG 3.7 target of reducing maternal deaths by two-thirds in 2030. In this regard, policy stakeholders should focus on creating an environment within households where women can independently make decisions about their own health needs and other private choices. Second, reflecting on the intersection of women's autonomy and urban-rural residency, policymakers need to pay key attention to sociocultural gendered norms in rural contexts. For example, it may be useful to implement culturally sensitive community-level educational programs on gender equality. Finally, in light of the impact of socioeconomic factors such as educational attainment and household income on women's mass media exposure to family planning, empowering women by addressing barriers to their access to formal education and employment opportunities may be useful for improving their exposure to these messages and improving their postpartum health outcomes.

References

Abada, T., & Tenkorang, E. Y. (2012). Women's autonomy and unintended pregnancies in the Philippines. *Journal of Biosocial Science, 44*(6), 703–718.

Ahmed, S., Li, Q., Liu, L., & Tsui, A. O. (2012). Maternal deaths averted by contraceptive use: An analysis of 172 countries. *The Lancet, 380*(9837), 111–125.

Ajaero, C. K., Odimegwu, C., Ajaero, I. D., & Nwachukwu, C. A. (2016). Access to mass media messages, and use of family planning in Nigeria: a spatial-demographic analysis from the 2013 DHS. *BMC Public Health, 16*(1), 1–10.

Alberti, K. P., Grellety, E., Lin, Y. C., Polonsky, J., Coppens, K., Encinas, L., et al. (2010). Violence against civilians and access to health care in North Kivu, Democratic Republic of Congo: Three cross-sectional surveys. *Conflict and Health, 4*(1), 2–7.

Alemayehu, M., & Meskele, M. (2017). Health care decision making autonomy of women from rural districts of Southern Ethiopia: A community based cross-sectional study. *International Journal of Women's Health, 9*, 213.

Amoyaw, J. A., Kuuire, V. Z., Boateng, G. O., Asare-Bediako, Y., & Ung, M. (2015). Conundrum of sexual decision making in marital relationships: Safer-sex knowledge, behavior, and attitudes of married women in Zambia. *The Journal of Sex Research, 52*(8), 868–877.

Anfaara, F. W., Atuoye, K. N., Antabe, R., Sano, Y., & Luginaah, I. (2020). Hepatitis B knowledge among women and men in the upper west region of Ghana: What sources of health information matter? *International Journal of Health Planning and Management, 35*(5), 1098–1110.

Antabe, R., Sano, Y., Anfaara, F. W., & Luginaah, I. (2021). Reducing HIV misconceptions among females and males in Malawi: Are we making progress? *AIDS Care, 33*(3), 408–412.

Bankole, A. S., & Westoff, C. F. (1996). *Mass media influences on contraceptive behavior and reproductive preferences*.

Basu, A. M. (1992). *Culture, the status of women, and demographic behaviour: Illustrated with the case of India*. Clarendon press.

Boateng, G. O., Kuuire, V. Z., Ung, M., Amoyaw, J. A., Armah, F. A., & Luginaah, I. (2014). Women's empowerment in the context of millennium development goal 3: A case study of married women in Ghana. *Social Indicators Research, 115*(1), 137–158.

Cleland, J., Conde-Agudelo, A., Peterson, H., Ross, J., & Tsui, A. (2012). Contraception and health. *The Lancet, 380*(9837), 149–156.

Corker, J. (2010). "Ligne Verte" toll-free hotline: Using cell phones to increase access to family planning information in the Democratic Republic of Congo. *Cases in Public Health Communication & Marketing, 4*, 23–37.

Democratic Republic of the Congo (DRC) Ministry of Health (MOH). (2014). Family Planning National Multisectoral Strategic Plan (2014–2020).

Development Media International. (2015). Evaluation of a family planning mass media campaign in Kinshasa. *Democratic Republic of the Congo*, 1–23. Retrieved from http://www.developmentmedia.net/uploads/9/0/0/3/90035669/kinshasa_family_planning_campaign_evaluation_report_final.pdf

Dyson, T., & Moore, M. (1983). On kinship structure, female autonomy, and demographic behavior in India. *Population and Development Review, 9*(1), 35–60.

Hernandez, J. H., Akilimali, P. Z., Muanda, M. F., Glover, A. L., & Bertrand, J. T. (2018). Evolution of a large-scale community-based contraceptive distribution program in Kinshasa, DRC based on process evaluation. *Global Health Science and Practice, 6*(4), 657–667.

Islam, M. M., & Hasan, A. S. (2000). Mass media exposure and its impact on family planning in Bangladesh. *Journal of Biosocial Science, 32*(4), 513–526.

Jacobs, J., Marino, M., Edelman, A., Jensen, J., & Darney, B. (2017). Mass media exposure and modern contraceptive use among married West African adolescents. *European Journal of Contraception and Reproductive Health Care, 22*(6), 439–449.

Jahan, Y., Rahman, S. A., Chowdhury, A. S., Chowdhury, S., Huq, K. E., & Moshiur, M. (2017). Trends in mass media exposure upon women: A review of Bangladesh demographic and health survey. *Global Journal of Medicine and Public Health, 6*(2), 1–13.

Kandala, N. B., Mandungu, T. P., Mbela, K., Nzita, K. P. D., Kalambayi, B. B., Kayembe, K. P., & Emina, J. B. O. (2014). Child mortality in the Democratic Republic of Congo: Cross-sectional evidence of the effect of geographic location and prolonged conflict from a national household survey. *BMC Public Health, 14*(1), 1–9.

Kansanga, M. M., Asumah Braimah, J., Antabe, R., Sano, Y., Kyeremeh, E., & Luginaah, I. (2018). Examining the association between exposure to mass media and health insurance enrolment in Ghana. *International Journal of Health Planning and Management, 33*(2), e531–e540.

Kayembe, P. K., Fatuma, A. B., Mapatano, M. A., & Mambu, T. (2006). Prevalence and determinants of the use of modern contraceptive methods in Kinshasa, Democratic Republic of Congo. *Contraception, 74*(5), 400–406.

Konkor, I., Sano, Y., Antabe, R., Kansanga, M., & Luginaah, I. (2019). Exposure to mass media family planning messages among post-delivery women in Nigeria: Testing the structural influence model of health communication. *European Journal of Contraception and Reproductive Health Care, 24*(1), 18–23.

Kwete, D., Binanga, A., Mukaba, T., Nemuandjare, T., Mbadu, M. F., Kyungu, M. T., et al. (2018). Family planning in The Democratic Republic of the Congo: Encouraging momentum, formidable challenges. *Global Health Science and Practice, 6*(1), 40–54.

Lawry, L. L., Barhobagayana, J., & Faye, O. (2019). Using a modified champion community approach for improving maternal, newborn, and child health outcomes in remote and insecure health zones in the Democratic Republic of Congo 2012–2017: a case study. *Journal of Global Health Reports, 3*, e2019068.

Lee, H., Park, S. J., Ndombi, G. O., & Nam, E. W. (2019). Community-based maternal and child health project on 4+ antenatal care in the Democratic Republic of Congo: A difference-in-differences analysis. *Reproductive Health, 16*(1), 1–10.

Mason, K. O. (1986). The status of women: Conceptual and methodological issues in demographic studies. In *Sociological forum* (Vol. 1, No. 2, pp. 284–300). Dordrecht: Kluwer Academic Publishers.

Menard, S. (2002). *Applied logistic regression analysis* (No. 106). Sage.

Mikatho, A., O'Connell, K.A., Ly, M., Kibungu, J., Garfinkel, D., Fofana Cisse, N., & Shalita. N. (2021). *Postpartum Family Planning (PPFP) in the Democratic Republic of the Congo: Results from a Key Intervention to Increase Counselling, Quality of Care, and Uptake among PPFP Clients.* Edited by A. Agarwal. Kinshasa: ExpandFP and Washington, DC: EngenderHealth.

Ministére de la santé publique. (2014). *Plan stratégique national à vision multisectorielle (2014–2016).*

Ministère du Plan et Suivi de la Mise en oeuvre de la Révolution de la Modernité (MPSMRM), Ministère de la Santé Publique (MSP) and ICF International. (2014). *Democratic Republic of Congo demographic and health survey 2013–14: Key findings*. MPSMRM, MSP et ICF International.

Muanda, M., Ndongo, P. G., Taub, L. D., & Bertrand, J. T. (2016). Barriers to modern contraceptive use in Kinshasa, DRC. *PLoS One, 11*(12), 1–13.

Mukaba, T., Binanga, A., Fohl, S., & Bertrand, J. T. (2015). Family planning policy environment in The Democratic Republic of the Congo: Levers of positive change and prospects for sustainability. *Global Health Science and Practice, 3*(2), 163–173.

Nigatu, D., Gebremariam, A., Abera, M., Setegn, T., & Deribe, K. (2014). Factors associated with women's autonomy regarding maternal and child health care utilization in Bale Zone: A community based cross-sectional study. *BMC Women's Health, 14*(1), 79.

Osotimehin, B. (2015). Family planning as a critical component of sustainable global development. *Global Health Action, 8*(1), 10–11.

Raudenbush, S. W., & Bryk, A. S. (2002). *Hierarchical linear models: Applications and data analysis methods* (Vol. 1). sage.

Sano, Y., Antabe, R., Atuoye, K. N., Hussey, L. K., Bayne, J., Galaa, S. Z., et al. (2016a). Persistent misconceptions about HIV transmission among males and females in Malawi. *BMC International Health and Human Rights, 16*(1), 16.

Sano, Y., Sedziafa, A. P., Amoyaw, J. A., Boateng, G. O., Kuuire, V. Z., Boamah, S., & Kwon, E. (2016b). Exploring the linkage between exposure to mass media and HIV testing among married women and men in Ghana. *AIDS Care, 28*(6), 684–688.

Sano, Y., Sedziafa, A. P., Vercillo, S., Antabe, R., & Luginaah, I. (2018a). Women's household decision-making autonomy and safer sex negotiation in Nigeria: An analysis of the Nigeria Demographic and Health Survey. *AIDS Care, 30*(2), 240–245.

Sano, Y., Antabe, R., Atuoye, K. N., Braimah, J. A., Galaa, S. Z., & Luginaah, I. (2018b). Married women's autonomy and post-delivery modern contraceptive use in the Democratic Republic of Congo. *BMC Women's Health, 18*(1), 1–7.

Starbird, E., Norton, M., & Marcus, R. (2016). Investing in family planning: Key to achieving the sustainable development goals. *Global Health, Science and Practice, 4*(2), 191–210.

Tran, N. T., Seuc, A., Tshikaya, B., Mutuale, M., Landoulsi, S., Kini, B., ... & Kulimba, D. M. (2020). Effectiveness of post-partum family planning interventions on contraceptive use and method mix at 1 year after childbirth in Kinshasa, DR Congo (Yam Daabo): a single-blind, cluster-randomised controlled trial. *The Lancet Global Health, 8*(3), e399–e410.

WHO, UNICEF, UNFPA, World BANK, & UN. (2019). *Trends in Maternal Mortality 2000 to 2017: Estimates by WHO, UNICEF, UNFPA, World Bank Group and the United Nations Population Division*. World Health Organization.

Woldemicael, G., & Tenkorang, E. Y. (2010). Women's autonomy and maternal health-seeking behavior in Ethiopia. *Maternal and Child Health Journal, 14*(6), 988–998.

World Health Organization. (2006). Report of a WHO Technical Consultation on Birth Spacing. No. WHO/RHR/07.1. World Health Organization, Geneva, Switzerland. https://apps.who.int/iris/bitstream/handle/10665/73710/RHR_policybrief_birthspacing_eng.pdf

Chapter 7
Researching Home While Studying Abroad: Navigating Positionality in Health Research in Rwanda

Germaine Tuyisenge, Yvonne Kasine, and Marie Paul Nisingizwe

Background

Conducting research in a familiar community comes with positionality ambiguities that researchers need to consider throughout all stages of the research process. Navigating positionality is crucial when conducting research in deprived settings or with vulnerable populations, especially due to power relationships between the researcher and the participants. Researchers who conduct research in their communities or with those they are very familiar with are cautioned to reflect on their positionality and how it impacts their interactions with the communities/participants to prevent harm. Thus, researchers should be equipped with the necessary tools to reflect on their positionality as they navigate relationships with their study communities.

The literature on positionality during fieldwork is largely by researchers from high-income countries (HICs) either in these countries or in low- and middle-income countries (LMICs). The literature on the positionality of non-westerners who conduct research in low- and middle-income settings is very little (Adu-Ampong & Adams, 2020; Chereni, 2014; Sharma, 2019). Thus, there is not much to refer to on positionality considerations while designing a study or conducting fieldwork in these settings. Many doctoral students from LMICs who are based in

G. Tuyisenge (✉)
London School of Hygiene and Tropical Medicine, London, UK

Y. Kasine
Arthur Labatt Family School of Nursing, Western University, London, ON, Canada
e-mail: ykasine2@uwo.ca

M. P. Nisingizwe
School of Population and Public Health, University of British Columbia, Vancouver, BC, Canada

J. A. Braimah et al. (eds.), *Health Geography in Sub-Saharan Africa*, Global Perspectives on Health Geography, https://doi.org/10.1007/978-3-031-37565-1_7

universities in western countries go back to do research in their home countries. Yet, there is little documentation on researchers' positionalities and ethical considerations regarding research in these contexts (Giwa, 2015).

Using examples from our experiences conducting health research in Rwanda, our home country, we reflect on positionality and its influence on our fieldwork as doctoral students studying in Canada. We discuss our roles as insiders to the communities we researched based on our familiarity with the country, yet outsiders to the topics of our respective research projects and from the fact that were living and studying outside the country for several years. The three of us were born and raised in Rwanda, where we obtained bachelor's degrees, and worked in different capacities within the country's health system. We are very familiar with the Rwandan culture and social norms, and we speak Kinyarwanda, our mother tongue and the official language and primary language of communication in Rwanda. In this chapter, we reflect on our positionality and what our social statuses did (or did not) mean to our participants and the influence on the research processes. We share experience from fieldwork while doing our respective projects to illustrate the contexts of positionality for each of us. We draw from three distinct studies that we conducted between 2017 and 2019 across different communities in Rwanda: The first study was a qualitative exploration of barriers to access maternal healthcare services in rural communities of Rwanda (Tuyisenge et al., 2019). In-depth interviews were conducted with new mothers who shared their experiences of accessing maternal health services with the researcher, GT, who did not have personal experience on motherhood thus reflects on positionality as an insider-outsider to this group. The second study was conducted by YK, a nurse-educator, who used in-depth interviews to explore nurses' experiences translating continuing professional development (CPD) to nursing practice in Rwanda (Kasine et al., 2018). YK shares experience on navigating power dynamics while building rapport with participants, some of them being former students and/or trainees on CPD programs. The third study, conducted by MPN, is an evaluation of a mass screening and treatment program for Hepatitis C Virus (HCV) in Rwanda (Nisingizwe et al., 2019). A retrospective study and in-depth interviews were conducted with individuals who have benefited from the HCV treatment through the program, some of them having completely healed while others were still on treatment. In this study, perceived responsibilities (of researcher) and expectations (of participants) are examined and their implications on data collection.

Positionality and Why It Matters: Insider, Outsider, or Both?

Defining the role of either an insider or outsider to the studied community is always challenging for many researchers who use qualitative interviews. The meanings accompanying such positionality bring researchers' attention to their roles and how these may impact the research process and outcomes (Giwa, 2015). Depending on the culture and community context, the identity dichotomy could be experienced when researching settings similar or different to those that researchers identify most with, such as rural vs urban, or when considering other characteristics of both the researcher and participants, such as age, marital status, religion, gender, education, etc. (Sultana, 2007). The intimate relationship between the researcher and participants could also be due to the shared culture and/or the researcher's familiarity with it (Isaacs, 2014). In some cases, researchers might identify themselves as insiders to the researched community based on the knowledge of the subject as a result of being immersed in the literature (King et al., 2018).

The researcher's positionality contributes to their ability to navigate researched communities and build trust with participants (Giwa, 2015). The following statement extracted from an interview with a couple who lost a child shows the essential role that positionality can play in data collection and how the participants may view the study and engage with the researcher:

> Because I know where you're sitting that you're interviewing from the glass house, cause when I said, "Has she ever had a personal loss, like a sibling?" and they [organization staff] said, "Well, not that we're aware of." And I said, "That's fine, it doesn't bother me, but, it's different if you're a member of the club." And I never want you to be a member of the club, don't get me wrong, but you talk differently to people who have lost somebody. (From Dwyer & Buckle, 2009, pp. 57)

Positionality is not negotiated with participants to agree on the role that the researcher occupies. Yet, its effect unfolds at different stages of the research depending on how each party considers their respective role in the research (researcher vs participant). As such, positionality may influence the interaction through what the participants shares and how they share it.

The preceding quote exemplifies what I encountered while researching maternal health with new mothers in Rwanda, something I had not personally experienced. Every single participant asked me if I was a mother. My answer often received follow-up questions, including questions on my age and my marital status (33 and single at the time). From there, some participants would criticize my choice of prioritizing studies over becoming a mother, on the one hand, because '*this is what is expected of you as a Rwandan woman,*' some would say; or '*nothing would ever fulfill you as motherhood,*' others told me. On the other hand, and sometimes to my surprise, a few participants congratulated me on '*my brave decision of following my heart*' and doing what is best for me because '*marriages are failing big time nowadays anyway*' and '*you cannot always rely on men to sustain your needs*' or '*Our government supports women empowerment,*' one or two participants emphasized, referring to the importance of education for women.

Such discussions came at different stages of the interview (before, during, or after). Interestingly, none of the participants went as far as asking me why I was in school and my plans after school. Maybe I could have justified my choices, I thought. Instead, every perspective I received was present and current, and the advice on what action to take was to be in the moment. As an insider of the Rwandan culture, I expected and respected such questions and took the time to answer to the participant's satisfaction, especially by listening to their advice. Participants' reactions to my situation were equally expected. Yet, my status as an outsider to motherhood was accentuated by being exposed to a different culture (Western) where such questions are considered sensitive and invasive. Thus, interactions triggered by the researcher's positionality as an insider or outsider to the studied community could, in some cases, make participants consider the researcher not necessarily as 'knowledgeable' about the topic.

Sanjari et al. (2014) highlight that depending on the level of sensitivity of the subject, participants might feel more comfortable sharing with an insider information they would not necessarily share with an outsider, which is stressed in the preceding quote. However, in some cases, this intimacy might prevent participants from fully expressing their views or feelings by assuming that the researcher knows what they are talking about (McDermid et al., 2014). In other situations, being an insider might raise questions about trust. For example, participants may conceal information when the researcher is considered a member of their 'club' who will know more about them and may judge their views or feelings. This could be encountered when participants want to protect their image. In this case, being an outsider might be advantageous to the researcher, as participants might open up more to a stranger who is interested in their stories and probably will not see them again.

It is important to acknowledge that there are advantages and disadvantages associated with either of the positions that the researcher might identify with. And no matter how much the researcher can delineate their research from personal experience on the topic, it is important to acknowledge that positionality as an insider or outsider affects to some degree how participants express themselves. Thus, there are always conflicting ideas about which position one occupies, and, in most situations, the researchers occupy the space in-between. However, according to Sultana (2007), those who go back home researching their communities cannot have complete membership of belonging within these communities because of the experiences and exposure from the global North that make them not entirely part of home communities anymore.

In fact, there is no wrong or right position. The most important thing is for the researcher to acknowledge their positionality and be aware of its implications at all the stages of the study to avoid any biases that might arise during the research. Knowing the position and how this might affect the research process is one of the critical issues the researcher needs to keep in mind (McGarry, 2016). And needless to say, researchers get more knowledgeable about the subject as they do more literature reviews on the subject and cannot be labelled as a complete outsider to their research topic. Thus, many researchers end up occupying the space between the two positions therefore they need to consider both roles rather than choosing one or the other.

Navigating Power Dynamics in a Hierarchal Society

Ethical considerations are important to navigate power dynamics appropriately while conducting research (Giwa, 2015). Discussions surrounding power relationships between the researcher and participants have been ongoing for the past two decades. Such debate cuts across several disciplines focusing on the impact of the researcher's identity and social position on participants. This discussion has also been relevant to qualitative health geographers for over a decade (Tarrant, 2014). Unsurprisingly, identity dilemma is one of the issues that researchers from LMICs face when they go back home to do research after spending some time in the HICs.

Sharing her experience while doing fieldwork in rural Bangladesh, Sultana (2007), a then-doctoral student in England, confessed the following:

> What perhaps concerned me the most about my positionality was the clear class difference: I was from the city, from an educated background, could read and write (in English no less!). Such overt differences immediately put me in a different location, and often in one of hierarchy, where people in rural areas have come to respect and be deferential to urban, educated elites. The fact that I wore shoes, a watch, carried a notebook, had a camera, all placed me in an irreconcilable position of difference. (Sultana, 2007, pp 378)

As noted by the author, such power differentials are experienced from the researcher's positionality, first as someone coming from the city to do research in a rural village. In addition, the fact that she was studying and living in a HIC broadened the power gap between her and the participants, who were in an impoverished rural area in a low-income country.

My study explored nurses' experiences translating continuing professional development (CPD) to nursing practice in Rwanda (Kasine et al., 2018). I interviewed 10 nurses who participated in a CPD program funded by a Western University-based project, the Maternal, Newborn and Child Health project in Rwanda (MNCHR). The same project funded my graduate studies in Canada, and, prior to that, I worked as a nurse educator in Rwanda. Thus, my positionality created power relations toward participants: first through my previous role with some of them as my former trainees; second through my relationship with the MNCHR project. I had to be mindful of such power dynamics, especially while building rapport with participants and introducing the research and its purpose, and throughout my interactions with participants during interviews.

A number of researchers have discussed ethical issues surrounding power dynamics while building rapport between the researcher and participants (Kornbluh, 2015; Le Dantec & Fox, 2015; McGarry, 2016). For example, the way the researchers introduce themselves and the study, how they communicate with participants, and how they show interest in their participants are among issues that help researchers narrow the power gap and support building rapport with participants (Le Dantec & Fox, 2015). Other forms of power relationships between the participant and researcher may be observed through how researchers encourage participation. This was particularly felt in my situation; since I was knowledgeable of the CPD content,

I had to create an environment where they felt comfortable sharing their experiences without the fear of being evaluated or judged.

As part of reducing the power gap, Taylor et al. (2014) call for negotiating and respecting the norms set for participation. They stress the importance of being respectful to participants if there are shifts based on their experiences from interviews. The authors explore the 'forcing participation' and 'bargaining participation' and remind researchers to put participants at ease while encouraging participation. In my case, my status as a researcher might have attenuated the power imbalances. However, it is important to note that power dynamics could go in two ways. For one, the participant is of low social status compared to the researcher (e.g., educated vs non-educated). For the other, the researcher is interviewing an elite participant, where they impose more control about the interview (Vorrath, 2013).

In some situations, the researcher might need to share their own experiences to narrow the power gap and make the participant feel at ease during the interview. This can create a feeling of increased trust among participants. Thus, a sense of friendship could develop through this rapport-building process. However, in some cases, the friendship might be faked from either side. Yet, faking friendship could affect access to other potential participants (Crow, 2013), thus reducing the trust with the researched community. Hence researchers are cautioned that the nature of the research conducted in contexts familiar to the researcher (such as community or cultures) and related power relationships may impact the quality of collected information (McDermid et al., 2014).

When building rapport and narrowing the power gap, researchers are urged to get close but not too close to the participants to avoid any misinterpretation of the relationship being built. Here it is important to emphasize that closeness is experienced differently depending on individuals and their cultural context. The researcher should be mindful of different aspects involving such relationships and be able to interpret them for mutual benefit. The researcher's priority is to ensure that the participant's current and future feelings are protected. Researchers are urged to know their place in the lives of their participants and be aware of how much impact every interview might have on the participants' lives.

Acknowledging Responsibilities and Managing Expectations: Do No Harm

The researcher's positionality should embody the do no harm responsibilities toward participants and should take into consideration their expectation to participate in the study. The nature of the study settings and the relationship between participants and gatekeepers (those who provide access to participants) may accentuate responsibility/expectation dynamics. This is, for example, when participants feel obliged to participate in the study or when they have false expectations of their involvement in the study (Tarrant, 2014; Taylor et al., 2014).

As highlighted by Ansoms (2013), research conducted in resource-poor settings often gives participants a feeling that the research is being undertaken with the purpose of 'doing something to making things better' or 'to have the participants' voices heard.' In this case, there is a certain level of expectation from the participant. It could be a direct benefit they expect from participating in the study, such as compensation, or an indirect benefit on the change that will come a as a result of the study.

In my research on evaluating a mass screening and treatment program for Hepatitis C Virus (HCV) in Rwanda, I had to reflect on my positionality toward the nature of the study and the participants and the level of vulnerability some of them might have as HCV patients. I was mindful of potential participant expectations that may result from their involvement in the study and based on existing relationship with gatekeepers (nurses responsible for the HCV program in hospitals who connected me with participants). The nurses followed up participants while on HCV medication; some were still on treatment while others had healed. During field work, I had to reassure participants and gatekeepers that participation in the study will not affect the healthcare–patient relationship and that their participation in the study will not influence any healthcare follow-up services as part of their HCV treatment.

In some cases, the researcher may be put in a vulnerable position in relation to the participant's expectation or knowing there is not much s/he can do when faced with situations that need support or action. This was the case when some of my participants revealed that they initially agreed to participate in the study, secretly hoping that it may have an impact on their HCV treatment. This is an example of how participants can expect future changes to their lives relative to the subject under investigation. And even though the optimum goal of research in most cases is to facilitate a positive change, it is important to keep in mind the extent to which the research can bring intended change; how far could those voices be heard or when and how much action could really be taken as a result of the study (Ferguson et al., 2017).

There is a constant recall that whatever could be done as a result of the research would probably take some time and that participants may not benefit directly (Ferguson et al., 2017). In this context, the author highlights the power the participants believe the research(er) has. This could have a significant impact in some communities, where for example, those not involved in the research may wonder why they are not involved and feel that those involved are getting more benefits (Ansoms, 2013). An example of how such a situation can happen is illustrated in the following quote on Ansom's experience while doing field work in Rwanda:

> One of our interviewees, for example, explicitly requested that the survey be conducted in the presence of his neighbour who was visiting him. He wanted to make sure that the neighbour could see what the survey entailed, and how minimal soap is as a reward". … "He told me that he was always afraid after meeting with a research team of being robbed as people in the neighbourhood suspected him of having received 'a lot of money.' (Ansoms, 2013, pp. 47)

In cases as the one highlighted in the preceding text, the researcher is put in a situation where s/he has to emphasize the role of the research and its benefits to participants. Here comes the notion of researcher responsibilities to manage participants' expectations. While conducting research, the researcher must be respectful of their values and the values of those being researched and their communities in general. It is possible that the researcher feels a certain level of responsibility toward the situation on the field, whether the participants expect them to do something about it or not. Another example from Ansoms reveals that:

> At one point, for example, I handed over a small amount of money to a family with a clearly very sick child that had to go to the hospital. However, during my next visit to this setting, I was confronted with a line of mothers asking for help for the urgent needs of their children.... But this left me with an awkward feeling of wondering whether being a researcher and being a decent human being were actually two different things. (Ansoms, 2013, pp. 48)

These statements emphasize the task of a researcher to protect participants. By not being able to help other mothers in the community, this researcher has probably changed the image of the mother who was supported in her community. Situations such as these are encountered by many researchers when the participants believe in some level of power that the researcher has (financial power, connections with local authorities, etc.). Hence researchers are called to clarify their roles to avoid any misinterpretations of their presence in the communities which can lead to false hopes. If not well managed, the expectations and responsibilities of the researcher could damage the relationship between the participants and the researcher, especially if the former feels the latter did not satisfy their needs but still wants to use them for their own benefit (Damme, 2013). The expectations from participants, as shown in the preceding text, could be of different nature, and so are the responsibilities of the researcher. Hence the reminder of the importance of providing participants with full information and explanation of the research prior to data collection and at all stages of research.

For example, Clark-Kazak (2013) stresses that interactions with participants during fieldwork can create a relationship with the researched having a feeling of doing a job for which they will be paid. Such a relationship is termed a "*donor–recipient*" relationship (Clark-Kazak, 2017, 2011). This is mostly experienced in impoverished populations or in settings where participants view the researcher as someone who can provide financial support, which can be the case for those who live in HICs doing research in LMICs, including students who go back home for research. This can have an impact on who agrees to participate in research and the kind of information they provide. This situation could also affect other studies that could be conducted in the future with the participants or within the communities as community members become resentful after some time of research involvement. To avoid the feeling of disappointment from both sides when responsibilities and expectations are not met, Van Damme (2013) advocates for designing studies that mutually benefit both sides (research and researched), and such benefit should be stated to participants at the beginning of the study and throughout the research process as needed.

Conclusion

The positionality ambiguities discussed here are some of the considerations that researchers must keep in mind during fieldwork to protect participants while collecting quality data. We have an obligation to provide useful information from our studies, and this should not be at the expense of participants' feelings. By following ethical norms on research involving humans, researchers have to be mindful of the blind spots surrounding their positionality and critically reflect on them at each stage of the research. Those who go back home to conduct research must be prepared for any potential issues related to their positionality and know how to manage them while keeping a flexible mindset for the unexpected. Hence, there are no standard procedures for navigating positionality dynamics during fieldwork, no matter how familiar one could be with the studied communities. Thus, each research should be considered unique in its nature and context, the researcher's identity, community settings, and background of participants.

References

Adu-Ampong, E. A., & Adams, E. A. (2020). "But you are also Ghanaian, you should know": Negotiating the insider–outsider research positionality in the fieldwork encounter. *Qualitative Inquiry, 26*(6), 583–592.

Ansoms, A. (2013). Dislodging power structures in rural Rwanda: From 'disaster tourist'to 'transfer gate'. Emotionaland ethical challenges for field research in Africa: The story behind the findings, 42–56.

Chereni, A.(2014). Positionality and collaboration during fieldwork: Insights from research with co-nationals living abroad. *Forum Qualitative Sozialforschung/Forum: Qualitative Social Research, 15*(3). https://doi.org/10.17169/fqs-15.3.2058

Clark-Kazak, C. R.(2011). *Recounting migration: Political narratives of Congolese young people in Uganda*. McGill-Queen's Press-MQUP.

Clark-Kazak, C. R. (2013). Research as 'Social Work'in Kampala? Managing Expectations, Compensation andRelationships in Research with Unassisted, Urban Refugees from the Democratic Republic of Congo. Emotionaland ethical challenges for field research in Africa: The story behind the findings, 96–106.

Clark-Kazak, C.(2017). Ethical considerations: Research with people in situations of forced migration. *Refuge: Canada's Journal on Refugees/Refuge: Revue Canadienne Sur Les Réfugiés, 33*(2), 11–17.

Crow, G. (2013). Going back to re-study communities: Challenges and opportunities. *Progress in Development Studies, 13*(4), 267–278.

Dwyer, S. C., & Buckle, J. L. (2009). The space between: On being an insider-outsider in qualitative research. *International journal of qualitative methods, 8*(1), 54–63.

Ferguson, L., Chan, S., Santelmann, M., & Tilt, B. (2017). Exploring participant motivations and expectations in a researcher-stakeholder engagement process: Willamette Water 2100. *Landscape and Urban Planning, 157*, 447–456.

Giwa, A. (2015). Insider/outsider issues for development researchers from the Global South. *Geography Compass, 9*(6), 316–326.

Isaacs, A. N. (2014). An overview of qualitative research methodology for public health researchers. *International Journal of Medicine and Public Health, 4*(4), 318.

Kasine, Y., Babenko-Mould, Y., & Regan, S. (2018). Translating continuing professional development education to nursing practice in Rwanda: Enhancing maternal and newborn health. *International Journal of Africa Nursing Sciences, 8*, 75–81.

King, N., Horrocks, C., & Brooks, J. (2018). *Interviews in qualitative research.* Sage.

Kornbluh, M. (2015). Combatting challenges to establishing trustworthiness in qualitative research. *Qualitative Research in Psychology, 12*(4), 397–414.

Le Dantec, C. A., & Fox, S. (2015). Strangers at the gate: Gaining access, building rapport, and co-constructing community-based research. *Proceedings of the 18th ACM conference on computer supported cooperative work & social computing*, 1348–1358.

McDermid, F., Peters, K., Jackson, D., & Daly, J. (2014). Conducting qualitative research in the context of pre-existing peer and collegial relationships. *Nurse Researcher, 21*(5), 28.

McGarry, O. (2016). Repositioning the research encounter: Exploring power dynamics and positionality in youth research. *International Journal of Social Research Methodology, 19*(3), 339–354.

Nisingizwe, M. P., Makuza, J. D., Serumondo, J., Nsanzimana, S., Hedt-Gauthier, B., Bansback, N., et al. (2019). The Cascade of Care for Hepatitis C Treatment in Rwanda: A Retrospective Cohort Study of the 2017–2019 Mass Screening and Treatment Campaign. *Viruses, 15*(3), 661–679.

Sanjari, M., Bahramnezhad, F., Fomani, F. K., Shoghi, M., & Cheraghi, M. A. (2014). Ethical challenges of researchers in qualitative studies: The necessity to develop a specific guideline. *Journal of Medical Ethics and History of Medicine, 7*, 14.

Sharma, S. (2019). Me again: Fieldwork, practice and returning. *Area, 51*(3), 508–515.

Sultana, F. (2007). Reflexivity, positionality and participatory ethics: Negotiating fieldwork dilemmas in international research. *ACME: An International Journal for Critical Geographies, 6*(3), 374–385.

Tarrant, A. (2014). Negotiating multiple positionalities in the interview setting: Researching across gender and generational boundaries. *The Professional Geographer, 66*(3), 493–500.

Taylor, R., Arvidson, M., Macmillan, R., Soteri-Proctor, A., & Teasdale, S. (2014). What's in it for us? Consent, access, and the meaning of research in a qualitative longitudinal study. In *Methodological challenges and new approaches to research in international development* (pp. 38–58). Springer.

Tuyisenge, G., Hategeka, C., Kasine, Y., Luginaah, I., Cechetto, D., & Rulisa, S. (2019). Mothers' perceptions and experiences of using maternal health-care services in Rwanda. *Women & Health, 59*(1), 68–84.

Van Damme, J. (2013). From scientific research to action in Southern Kivu: Ethical dilemmas and practical challenges. In *Emotional and ethical challenges for field research in Africa* (pp. 84–95). Springer.

Vorrath, J. (2013). Challenges of interviewing political elites: A view from the top in post-war Burundi. In *Emotional and ethical challenges for field research in Africa* (pp. 57–69). Springer.

Part III
Environment and Health

Chapter 8
Water, Sanitation, and Hygiene Interventions for COVID-19 and the Health of Vulnerable Populations in Sub-Saharan Africa

Sarah L. Smiley, Ellis A. Adams, Benjamin D. Agbemor, and Hilary Hungerford

Introduction

The emergence and spread of COVID-19 highlighted the importance of access to water, sanitation, and hygiene (WASH). Health officials stressed the importance of regular handwashing with soap and water for COVID-19 control and prevention. However, billions of people around the world struggle to meet their daily WASH needs. Efforts by the United Nations (UN) to rally global support toward improving access to WASH began in the 1970s with the Mar De Plata Conference on Water. Subsequently, the UN launched the International Drinking Water Supply and Sanitation Decade. Lessons from the first water decade informed the UN's successive initiatives, such as the Millennium Development Goals and the Sustainable Development Goals (SDGs) which have focused on addressing the WASH needs of the world's poorest population. Notwithstanding these efforts, access to WASH services remains lower in sub-Saharan Africa (SSA) than in other world regions, and substantial inequalities exist within African countries.

Progress toward the SDG of universal and equitable access to safe and affordable drinking water for all is measured by the proportion of the population using safely

S. L. Smiley (✉) · B. D. Agbemor
Department of Geography, Kent State University, Kent, OH, USA
e-mail: ssmiley8@kent.edu; bagbemor@kent.edu

E. A. Adams
Keough School of Global Affairs, Eck Institute for Global Health, University of Notre Dame, Notre Dame, IN, USA
e-mail: eadams7@nd.edu

H. Hungerford
Department of Earth Science, Utah Valley University, Orem, UT, USA
e-mail: hilary.hungerford@uvu.edu

J. A. Braimah et al. (eds.), *Health Geography in Sub-Saharan Africa*, Global Perspectives on Health Geography, https://doi.org/10.1007/978-3-031-37565-1_8

managed water services. Safely managed water is an improved source located on premises, available when needed, and free from contamination. In 2017, 71% of the world's population used a safely managed source but just 27% of SSA's population did so (UNICEF & WHO, 2019). Progress toward the goal in the region is most constrained by the low numbers of on-premises water sources. Not only does the region lag behind the world, but it also has gaps between populations. Just 12% of its rural population has access to a safely managed water source while 50% of its urban population does (UNICEF & WHO, 2019).

The SDGs also seek adequate and equitable access to hygiene for all. Progress is measured by the proportion of the population with on-premises access to a handwashing facility with soap and water. In 2017, 25% of SSA's population had access compared with 60% globally. Gaps exist within the region, with 17% of Africa's rural population and 37% of its urban population having access to a facility with soap and water (UNICEF & WHO, 2019). Especially troubling is that 41% of the region's population lacks access to any type of handwashing facility (UNICEF & WHO, 2019). Wolf et al. (2019) found that people with access to a designated handwashing facility are more than twice as likely to practice handwashing with soap than those without a facility. Although that research was regarding potential fecal contact, it demonstrates the real health benefits of a handwashing facility for COVID-19 prevention too.

COVID-19 had a disproportionate impact on populations in SSA that have limited access to WASH facilities and pre-existing vulnerabilities. From poor urban residents to migrant and refugee populations to women and girls, vulnerable populations face critical challenges that COVID-19 worsened. This chapter identifies the ways in which the lack of access to water, sanitation, and hygiene (defined here as WASH insecurity) potentially increased the impact of COVID-19 on vulnerable populations in SSA. The chapter also identifies why well-intentioned WASH interventions to control COVID-19 in the region failed to address the specific needs of poor, under-resourced, and vulnerable populations. The chapter begins with an overview of WASH insecurities among vulnerable populations, presents case studies of different state responses and interventions to COVID-19, and reflects on the policy and structural weaknesses that undermined these interventions.

Vulnerability and WASH

The COVID-19 pandemic disproportionately affected vulnerable households and populations in SSA. There is growing concern that the pandemic and its associated preventive and control measures increased vulnerability and risk, affected livelihoods, and ultimately compromised health and well-being. Recent observations hold that the pandemic deepened these vulnerabilities and impacts because the region was already mired in widespread poverty and income inequalities; poor water, sanitation, and hygiene access; and poor health care (Okoi & Bwawa, 2020).

Urban Slums and Informal Settlements

More than 50% of SSA's urban population lives in slums and informal settlements where water insecurity is pervasive and undermines health and well-being (Adams et al., 2020). In such underserved areas, the chronic lack of safe water combined with poor housing, outdated infrastructure, and socio-relational and institutional failures exacerbate the impact of COVID-19. Across the region, slums and informal settlements are unplanned and overcrowded, lack basic amenities, and have subpar housing. These slums house an overwhelmingly large proportion of residents who are poor and do not have access to household taps and therefore rely on public standpipes; sachet water; and resold water from water tankers, vendors, and private wells (Adams & Vásquez, 2019). For those households lacking household taps, the necessity of fetching water from public spaces presented a risk for COVID-19 exposure and infection. Unless they were able to pay higher prices for delivered water, these households were unable to avoid entering crowded spaces.

A rapidly growing set of studies have reflected on how the COVID-19 pandemic affected slums and informal settlements given the vulnerabilities they face even without a global health crisis. One emerging concern was that urban slums could be a potential source of COVID-19 infection and spread because of both their sheer density and the near impossibility for households to adhere to social distancing and lockdowns (Ilesanmi et al., 2020). Another area of concern in recent studies was that lockdowns, physical distancing, and other COVID-19 control mechanisms unduly affected slum residents because many work in the informal sector and households predominantly use public water and toilet facilities (Chirisa et al., 2020). A long chain of water delivery from the point of fetching, transportation, and use at the household is known to increase water contamination in slums. There were concerns that the same pathways to water contamination potentially facilitated the spread of COVID-19 in slums (Haddout et al., 2020).

Migrant and Refugee Populations

SSA currently hosts 26% of the world's refugee population as numbers in recent decades soared from multiple, longstanding conflicts and crises in Nigeria, Central African Republic, and South Sudan (UNHCR, 2021). COVID-19 renewed health and well-being concerns about the millions of refugees in the region. Among the many challenges and unhealthy conditions that Africa's refugee and displaced populations face, many refugee camps possibly became hot spots for COVID-19 spread because of the lack of access to safe water, poor sanitary conditions, substandard living conditions, and overcrowded environments that make handwashing and social isolation near impossible (Okoi & Bwawa, 2020). Evidence from Uganda shows that the pandemic increased the susceptibility of refugee populations to stigma, gender-based violence, and anxiety (Bukuluki et al., 2020). Previous work shows that the precarious physical and environmental conditions in refugee camps contribute regularly to diarrhea and cholera outbreaks (Bwire et al., 2021). These and many other existing conditions among refugees such as hunger, poverty, poor sanitation, and food insecurity undermine refugee livelihoods, not to mention that handwashing and social distancing interventions were harder to adhere to in refugee camps. In South Africa, for example, COVID-19 exacerbated already existing vulnerabilities faced by refugees and undocumented migrants, especially women (Mukumbang et al., 2020).

Women, Girls, and Pregnant and Postpartum Women

Poor access to water and sanitation services worsens the health and livelihood of women and girls in SSA, a region with half of the global population without access to improved water sources and nearly 700 million without access to improved sanitation facilities (UNICEF & WHO, 2019). Over two-thirds of the region's population must leave their premises to collect water and this task falls disproportionately on an estimated 3.36 million girls and 13.5 million women (Graham et al., 2016). Because the burden of water collection, household sanitation, and family welfare falls disproportionately on women and girls in the region, responsibilities that are often shaped by deeply entrenched cultural norms and expectations, COVID-19 had a relatively more devastating impact on them (Adams et al., 2021). COVID-19 increased water demand and use globally because of handwashing prescriptions to control infections and spread. In settings where access to water entails walking to a source rather than using a household tap, women and girls can come under intense pressure from their families to satisfy household water needs. Aside from taking care of family WASH needs, women and girls must take care of their own reproductive and menstrual needs such as regularly washing clothes and underwear, especially when sanitary pads are either scarce or too costly and where reusable menstrual cloths are the cultural norm. Women and girls may feel added pressure to defer their own water needs at the cost of dehydration, poor menstrual health, and poor personal hygiene (Collins et al., 2019). Lockdowns likely imposed additional mental health burdens on women because family expectations on water fetching and cleaning will remain. Other observers raised deep concerns that nearly 2.5 million girls were at risk of child marriage because of the pandemic (Cousins, 2020), a disproportionately high number of whom were from SSA and South Asia.

African Responses to COVID-19

The first case of COVID-19 on the African continent was diagnosed in Egypt on February 14, 2020. After this initial case, the number of cases on the continent, and especially in SSA countries, has remained low with the exception of South Africa.

Different reasons have been put forward for these low levels including the region's experience with previous health pandemics such as HIV/AIDS and Ebola, its youthful population, and its quick response to implement restrictions. Countries around the world responded to the pandemic with public health orders such as mask mandates, capacity restrictions, travel restrictions, curfews, and closures.

Yet a significant difference between SSA and other countries was the response speed. Many countries in SSA responded before a significant number of cases were detected within their borders. For example, Ivory Coast began enhanced airport screenings on January 2, 2020, but it did not record its first case until March 11 of that year (Massinga Loembé et al., 2020). Across the African continent, countries used both containment and mitigation such as travel bans; mandatory quarantines; closed borders; and restrictions on movement, public gatherings, and schools. In fact, by March 20, 2020, despite there being less than 100 COVID-19 cases in the African continent, countries were imposing lockdowns and curfews (Massinga Loembé et al., 2020).

As Karim et al. (2021) explain, Rwanda implemented unique, aggressive, and successful COVID-19 responses all guided by a six-month National Preparedness and Response Plan. It established a toll-free hotline to report symptoms, used social media to counter misinformation, used drones to disseminate information, and used robots to monitor patients in hospitals. The country's Ministry of Health began issuing infection controls 1 week after the first case of COVID-19 was detected in Asia and more than 1 month before the first case was detected in Rwanda. After restrictions were eased, the government began random street testing campaigns in Kigali. Not all countries took the same level of action as Rwanda. For example, Tanzania last released COVID-19 health data, including the numbers of deaths and infections, on April 29, 2020, and on June 8, 2020, it declared the disease had been eliminated entirely from within its borders (BBC, 2020).

Those containment and mitigation measures, while necessary to slow the spread of the disease, had serious economic implications on incomes and employment rates. Some governments took direct economic action through payments and increased unemployment assistance. For example, Kenya implemented direct cash transfers via mobile banking to the most vulnerable residents of informal settlements to help them meet their basic needs (Gentilini et al., 2020). Other governments took more indirect action by limiting evictions, waiving late fees, and providing free electricity. In Ethiopia's Tigray State, evictions were halted and rents reduced by half (Gentilini et al., 2020).

Some countries in SSA implemented temporary WASH interventions to contain COVID-19. These interventions recognized the central role that water access and handwashing play in stopping the pandemic and sought to relieve some of the economic burdens of other responses. Similar to previous WASH interventions, these COVID-19 responses did not necessarily reach the most vulnerable and water insecure households (Smiley et al., 2020). Although many different WASH interventions were implemented, from installing handwashing stations in public spaces to subsidizing household water supplies to providing free household water supplies, they were usually rolled out without careful consideration for the specific needs of poor households.

There was a clear rationale for using WASH interventions to fight COVID-19. A main strategy of disease control and prevention is regular hand washing so ensuring households have enough water each day was essential. In fact, research from Uganda found both increases in handwashing frequency and in the quantity of water used for handwashing after March 2020 (Sempewo et al., 2021). Increased costs from this behavior change, however, may have led some water users to compromise on hygiene practices. Providing free or subsidized water relieved some economic pressure on households who were either bringing in reduced or no incomes due to lockdowns, curfews, and business closures. This provision may also have helped ensure water users did not compromise on hygiene because of water costs.

Case Studies of COVID-19 WASH Interventions in Africa

This section provides case studies from the Democratic Republic of Congo, Ghana, Niger, and Uganda. Each country provided WASH interventions to its residents although the amounts, time periods, and number of beneficiaries varied. Policy concerns specific to each country are also discussed.

The Democratic Republic of Congo

In the Democratic Republic of Congo, one response to the COVID-19 pandemic was to provide free water during the months of April and May 2020 for everyone with direct access to the piped water network of *Regie de Distribution d'Eau* (REGIDESO; Direction of Water Distribution), which serves mostly urban areas across the country (Olukya, 2020). The REGIDESO network is the largest network in Kinshasa, but in peri-urban areas there also exist a number of other water access points, including public REGIDESO water taps and water cooperative networks such as *Associations d'Usagers de Réseaux d'Eau Potable* (Association of Drinking Water Network Users). While 70% of Kinshasa's 14 million residents access water through these various networks, many of them do not have direct household connections (World Bank, 2021). As is the case in many African cities, water access and connection data in Kinshasa is undependable, with REGIDESO numbers often inflated to meet reporting goals (Bédécarrats et al., 2019). Bédécarrats et al. (2019) estimate that the population with direct access to the REGIDESO network in Kinshasa is close to 50%, with declining rates as you move toward the outskirts of the city. In peri-urban neighborhoods, households spend an average of 28 minutes fetching clean water, compared with 3–7 minutes in central and suburban Kinshasa (Bédécarrats et al., 2019).

Kinshasa is one of Africa's most densely populated cities (World Bank, 2021) and the water network is weak and patchy at best. While covering households' water bills for the first few months of the COVID-19 pandemic undoubtedly helped many households, inflation and poverty continued to rise, and no further water payment relief was apparent after a couple of months. As a result, many international organizations increased funding to water projects in Kinshasa, including the World Bank and the International Committee of the Red Cross. One of those World Bank funded projects was the country's Strategic Preparedness and Response Project which brought WASH services to healthcare facilities (World Bank, 2020). Another project was the Urban Water Supply Project which financed 20 shared water points in densely populated areas lacking water access. The water at these points came from tanker trucks, and the goal was that the infrastructure will remain after the pandemic (World Bank, 2020).

In Kinshasa, the 2 months of free water had a number of unforeseen circumstances. First, water use increased across the city leaving some neighborhoods without water for months on end (Bossay, 2020). These shortages occurred despite the water treatment stations working at full capacity across the city (Sopi, 2020). For example, residents of Ngafani district, which is located in central Kinshasa, already struggled with access to water before the COVID-19 pandemic and with increased water use they reported water scarcity and unreliability (Bossay, 2020). Second, most residents in peri-urban Kinshasa did not have piped water access and thus relied on water points far from their residences, but these water points were not included in the free water initiative (ICRC, 2020). Since water consumption increased because of more frequent handwashing, both the amount paid for water and the time spent getting water also increased. Since residents believed that water used for handwashing as COVID-19 prevention could not be safely recycled like other grey water, they needed even more water than before and made more trips to fetch it (ICRC, 2020). Concerns about this free water policy were not just raised in Kinshasa. Reports from Goma stated that water bills doubled for the months of June and July to recoup lost revenue from the 2 months of free water (PERC, 2020).

Ghana

Ghana implemented its free water initiative in phases and covered the 15-month period from April 2020 to June 2021. After the first phase expired in June 2020, extensions were announced in July 2020, January 2021, and March 2021. In the initial phases, water was free for all domestic customers, but the later phases restricted free water to those domestic consumers who used less than five cubic meters of water per month. This limit on quantity was to check the misuse of water, a situation that was observed during the initial phases. If multiple households shared a water meter, they could collectively exceed the five cubic meters of water per month and may not have enjoyed the free water initiative (Cooper, 2020).

Even before the free water initiative was announced, the government had begun a vigorous hand hygiene campaign. Ghanaians were encouraged to regularly wash their hands with soap and running water both at home and in public areas. Frequent use of alcohol-based hand sanitizers was encouraged in cases where it is difficult to wash hands. Managers of public places such as transport terminals, shared toilets, and offices were directed to provide handwashing stations. Observations suggest, however, that these facilities were used infrequently (Bonful et al., 2020; Fielmua et al., 2021).

The WASH-related interventions implemented during the pandemic had both successes and challenges. For some domestic customers, the free water program was beneficial, but this benefit only lasted while the taps were running. As was the case with Ghana's water supply prior to the implementation of the initiative, there were reports of interruptions in water availability during the pandemic. Unfortunately, the government did not roll out a corresponding state-wide expansion, repair, or maintenance program alongside the directives for free water provision. Thus, water users who had previously been relying on vendors due to a lack of access to formal water systems continued to use such services and pay for them. The irregular flow of the free water also compelled some beneficiaries to resort to the occasional use of water from unsafe sources. Another major omission of Ghana's free water directive was the non-inclusion of rural communities that rely on handpump services. In some of these communities, user fees are paid toward the cost of operations and maintenance (Smiley et al., 2020). It was impractical to provide reimbursement for water consumption in these handpump communities because of the nature of the water supply technology and the arrangements for their operations. However, a directive for another form of support – such as government-funded repairs and maintenance – would have been a suitable alternative to the free water initiative.

Reimbursements to service providers for water supplied for free was slow. As of January 2021, when a new phase of the initiative was announced, no public or private service providers had received full repayment for water supplied from July to December. Some providers had outstanding payments from the first phase while others had received as little as 4% of the total amount requested (USAID, 2021). This slow reimbursement was contrary to the government's original schedule of monthly repayments and affected the water production capacity of some water systems. For example, the independent providers who could not cope with these delays reverted to their prior arrangement of collecting user fees from customers. The delayed payment of reimbursements reportedly led to the depletion of the operational reserves for some water systems. The same challenge resulted in other private and community-based water systems deferring payments for critical expenses such as staff salaries or scheduled maintenance activities. Having a situation of low financial reserves immediately following a period of intense use of water supply systems could have affected water systems' sustainability in the future (USAID, 2021). Should the anticipated adverse effects of partial or complete system shutdown occur, it will lead to further interruptions in the water supply, and the poor who already have low adaptive capacity to water insecurity would suffer the most.

Niger

Niger has some of the world's highest population rates and faces a number of challenges with access to water (Goujon et al., 2020). Niamey, the capital and largest city, also experiences high growth rates, with an average annual growth rate of 5% (UN, 2018). The national and city governments have been unable to keep pace with such rapid growth and consequently, urban services are fragmented, and there is unequal access for residents. Although great progress has been made in the last 10 years in expanding networked water services across the city through additional water treatment stations and subsidized water connections, nearly 40% of Niamey's residents remain without reliable access to a piped water network (Keough & Youngstedt, 2019).

In late March 2020, the government of Niger announced a number of COVID-19 relief measures, including the payment of water bills for the months of April and May (IMF, 2021). The *Société d'Exploitation des Eaux du Niger* (Niger Water Exploitation Company) issued a communique that explained the state would cover the first 10 cubic meters, which they refer to as the subsidized or social welfare portion of the pricing system. Networked water in Niger follows a tiered pricing model, with higher prices for higher consumption amounts. The first tier, which is the least expensive water, is up to 10 cubic meters of water use. This amount was automatically deducted from consumers' bills for the months of April and May 2020. For the estimated 50–70% of households in Niamey with direct water pipes to their homes (Bontianti et al., 2014; Keough & Youngstedt, 2019), this subsidy likely helped many households save approximately US $3. Although that amount was substantial savings for households faced with poverty and high living expenses, households with the highest poverty levels were unlikely to have water taps in their residences. Moreover, households with more residents use substantially more than 10 cubic meters of water a month; generally, the upper middle class and elite have smaller household sizes than poorer households (Bontianti et al., 2014).

People without access to water at their residences rely on a variety of water systems, including having water taps with pre-paid meters, using small-scale water vendors who deliver jugs of piped water to households, getting water from public standpipes, buying bottled water, relying on neighbors, and occasionally using wells (Keough & Youngstedt, 2014, 2019). For public water taps (called *bornefontaines*), water was free for the months of April and May. Households reasonably close to public water taps could access water for free if they traveled to the tap directly. Many households, however, use small-scale water vendors because the distance to the tap is too far or they face other barriers to mobility. These small-scale vendors usually get water from public water taps but charged a delivery fee even when the water itself was free.

Another concern was that April and May are the hottest months of the year in Niamey, with average daily temperatures of over 100 °F, and are consequently among the months with the highest water use. Taps regularly run dry during the hot season because of the increased demand for water and increased demand for electricity that fuels the water pumping stations (Hungerford, 2012). Households regularly reported that during the hot season they would access water at night or early morning when demand across the city was lower and they were more likely to have water at public taps (Hungerford, 2012). During the COVID-19 pandemic, however, a national curfew was imposed which left households unable to access water during non-peak times.

Uganda

Uganda recorded its first case of COVID-19 on March 21, 2020 (Olum & Bongomin, 2020). WASH-related interventions targeted the vulnerable segments of the population and areas with high chances of the virus spreading. Some of the relief efforts were centrally planned and coordinated through the Ministry of Water and Environment and the National Water and Sewerage Corporation (NWSC). In the national capital, for example, the NWSC together with the Kampala Capital City Authorities provided some 300 water points in public spaces such as markets and transport terminals (Bukenya, 2020).

In urban and informal settlements, it was recognized that COVID-19 safety protocols such as regular handwashing would be difficult to observe in the absence of a clean water supply. Some of these settlements relied on water supplied in portable water containers by vendors on motorbikes. However, the operations of these vendors were disrupted by the nationwide restrictions on movements (Danilenko, 2020). In response, the NWSC partnered with corporate entities such as MTN Uganda to extend temporary water access to residents in water-stressed areas in Kampala and other large urban communities (Bukenya, 2020). These previously unserved areas were provided with high-capacity water storage tanks and emergency water reservoirs which were regularly refilled using trucks (Bukenya, 2020; Danilenko, 2020). Water at the public water points and at these storage tanks was sold at subsidized rates (Bukenya, 2020).

The idea of free water access for Ugandans was proposed by some members of Parliament but never enacted (The Independent, 2020). Meanwhile, the NWSC suspended the disconnection of water services for customers who did not pay their bills and subsidized water that was delivered through public standpipes (Bukenya, 2020; Cooper, 2020; Danilenko, 2020). These steps could be described as a compromise in the absence of a water bill waiver from the government, especially for people who had their means of livelihoods disrupted and would have required some form of external support to keep their taps running.

Among the development partners working in Uganda, UNICEF was particularly visible in providing WASH-related interventions in communities and institutions such as schools and healthcare facilities. These interventions included the rehabilitation of boreholes in selected districts, the construction of a piped water system, and the donation of WASH supplies such as soaps and sanitizers to schools and healthcare facilities (Ntabadde, 2021). Although the contributions by UNICEF and other development partners were commendable, they were limited in scope. Smith et al. (2020) noted that the temporary ban on public transport meant that handpump mechanics were greatly constrained in attending to communities that may need their services.

Limitations of WASH Interventions

The provision of free or subsidized water during a public health pandemic was admirable and well-intentioned. The health benefits from the increased ability to wash hands are clear. Prüss-Ustün et al. (2014) estimated that in 2012, 842,000 deaths from diarrhea were attributed to WASH, including nearly 300,000 from inadequate hand hygiene. Handwashing with soap reduces incidences of diarrhea (Cairncross et al., 2010; Prüss-Ustün et al., 2014), reduces the risk of respiratory infection (Mbakaya et al., 2017; Rabie & Curtis, 2006), and provides protection against pandemic influenza (Saunders-Hastings et al., 2017). Although these health benefits may be clear, there is no guarantee that all interventions will succeed. Prior to the COVID-19 pandemic, billions of dollars were spent each year by the global community on interventions designed to increase access to safe water, including both shared sources and private household connections (WHO, 2017). Yet many of these interventions underperform, and 30–50% of WASH projects fail after 2–5 years (UNDP-SIWI Water Governance Facility & UNICEF, 2015).

The complexity of water provision in terms of source types and locations across SSA makes successful interventions difficult, and it is inevitable that some populations and locations will benefit more than others. Free or subsidized water policies were likely most beneficial for homes with piped and metered connections from a public water utility provider. In those cases, as seen in the Ghana and Niger case studies, the public utility was able to directly credit water bills. Yet the reality across SSA is that most households lack a direct piped connection. In 2017, just 27% of people in the region had water accessible on premises, and those on-premises sources can include both wells and utility-provided water (UNICEF & WHO, 2019). Wealthier, privileged households are more likely to have on-premises water which meant that more vulnerable households – including ones with housing and water insecurity – were overlooked.

Rather than direct piped connections, many households across SSA purchase resold water from alternative sources such as water tankers, pushcart vendors, public standpipes, neighbors, and private wells. These alternative sources rely on a long chain of suppliers that may be formal or informal or a mix of the two. Providing free or subsidized water to those customers who pay for water by the bucket was difficult for several reasons. Informal providers may have had difficulty complying with free water policies, especially if they lacked water meters but were required to submit detailed data for reimbursement. They may not have trusted public institutions to provide full or timely reimbursement for their up-front costs. Likewise, informal sellers may have been unwilling to provide free water since they relied on revenue to pay resellers and fund source maintenance. These concerns were founded. Data from Ghana show that more than two-thirds of interviewed water users were still paying for water despite the government directive for free domestic water (USAID, 2021).

In addition to favoring those wealthier households with direct connections, these WASH interventions also favored urban areas. In Ghana, for example, the free water directive did not explicitly cover rural areas where handpumps were the major source of potable water which meant that some providers continued to charge for water. These rural areas were excluded since it would have been difficult to provide reimbursement to all handpump communities given that water was not metered from these facilities. Likewise, the system used for tracking the volume of water distributed to domestic customers in urban areas meant that the supply company was able to capture and submit accurate data on free water to ensure prompt processing of repayments. Piped systems operating in small towns, however, faced challenges in submitting data on the volume of water provided for free. Underlying issues such as a lack of bulk production or distribution meters in some cases, and inadequate logistics meant that data on the volume of water distributed for free was estimated in some cases or ridden with errors in other instances (Smiley et al., 2020). Ultimately, these factors contributed to the delayed submission of free water data and contributed to the slow pace of reimbursements.

In extreme cases, the complications with data capture and submission, together with delayed reimbursement and low financial reserves, compelled some community-managed piped systems in small towns or rural areas to operate below their full capacity or demand user fees to keep the water systems working. Even with short-term measures such as the installation of high-volume water storage tanks and the deployment of tanker services, urban areas were favored over rural communities; these emergency response measures were more convenient and practical in urban areas and had a larger impact by reaching more people. Free or subsidized water did not necessarily alleviate all costs incurred by households. In communities where water is already provided for free, households are instead expected to contribute to construction and maintenance costs for water sources (Adams & Smiley, 2018). Free or subsidized water policies addressed only the water cost itself and not other related – and potentially significant – expenses.

Conclusion

The first case of COVID-19 was detected on the African continent in Egypt on February 14, 2020. Even before that first diagnosis, many SSA countries took proactive steps to minimize the impacts of the pandemic. As a result, infections and deaths remained relatively lower for this region than in other world regions. Various explanations were offered for these lower incidences including past experiences with pandemics and early restrictions. Still, concerns remained that perhaps low testing capacities and undercounting also played a role in low reported infection rates. The restrictions imposed during the pandemic including the closing of businesses and borders, airport screenings, and lockdowns played a role in mitigating the health crisis. However, these restrictions also had serious economic impacts on households.

SSA's vulnerable populations faced significant challenges in trying to adhere to preventive and control measures to contain COVID-19, especially considering that the region trails (along with South Asia) other regions on all three components of WASH – water, sanitation, and hygiene. Although many scholars and practitioners concede that global reports from the WHO and UNICEF underestimate the burden of WASH, SSA nonetheless constitutes nearly half of this burden. The WASH challenge in the region is longstanding and severe and it interacted with the pandemic to deepen the plight of women and girls, refugee populations, those that live in urban slums, and many other vulnerable groups outside the scope of this chapter.

To counter these negative, and unintended consequences, governments implemented various forms of social relief. One such program observed in several countries in SSA was the provision of free or subsidized water. Although these WASH interventions took different forms – from unlimited domestic water in Ghana to limited amounts for only those customers with piped connections in Niger and the Democratic Republic of Congo – they undoubtedly had benefits in promoting handwashing and helping to offset reduced incomes. Yet the question was whether these interventions reached the most vulnerable and water-insecure households. By privileging those households with direct household connections – whether through explicit policy or through slow reimbursements that made informal providers reluctant to follow directives – the poorest people who rely on informally resold water were overlooked.

The four case studies included here provide lessons for designing effective health interventions in future pandemics or emergencies. To ensure effectiveness, critical reflection on gaps in policies, adequacies of water quantity, and insufficient time periods of relief are necessary to shape future interventions. One necessary component of responses to pandemics such as COVID-19 or even natural disasters is to ensure that policies are inclusive and approach water security holistically (Anim & Ofori-Asenso, 2020).

References

Adams, E. A., & Smiley, S. L. (2018). Urban-rural water access inequalities in Malawi: Implications for monitoring the Sustainable Development Goals. *Natural Resources Forum, 42*, 217–226. https://doi.org/10.1111/1477-8947.12150

Adams, E. A., & Vásquez, W. F. (2019). Do the urban poor want household taps? Community preferences and willingness to pay for household taps in Accra, Ghana. *Journal of Environmental Management, 247*, 570–579.

Adams, E. A., Stoler, J., & Adams, Y. (2020). Water insecurity and urban poverty in the Global South: Implications for health and human biology. *American Journal of Human Biology, 32*(1), e23368.

Adams, E. A., Adams, Y. J., & Koki, C. (2021). Water, sanitation, and hygiene (WASH) insecurity will exacerbate the toll of COVID-19 on women and girls in low-income countries. *Sustainability: Science, Practice and Policy, 17*(1), 86–90. https://doi.org/10.1080/15487733.2021.1875682

Anim, D. O., & Ofori-Asenso, R. (2020). Water scarcity and COVID-19 in sub-Saharan Africa. *The Journal of Infection, 81*(2), e108–e109. https://doi.org/10.1016/j.jinf.2020.05.032

BBC. (2020, June 20). *Coronavirus: John Magufuli declares Tanzania free of Covid-19*. BBC. https://www.bbc.com/news/world-africa-52966016

Bédécarrats, F., Lafuente-Sampietro, O., Leménager, M., & Sowa, D. L. (2019). Building commons to cope with chaotic urbanization? Performance and sustainability of decentralized water services in the outskirts of Kinshasa. *Journal of Hydrology, 573*, 1096–1108.

Bonful, H., Addo-Lartey, A., Aheto, J., Ganle, J., Sarfo, B., & Aryeetey, R. (2020). Limiting spread of COVID-19 in Ghana: Compliance audit of selected transportation stations in the Greater Accra region of Ghana. *PLoS One, 15*(9), e0238971. https://doi.org/10.1371/journal.pone.0238971

Bontianti, A., Hungerford, H., Younsa, H. H., & Noma, A. (2014). Fluid experiences: Comparing local adaptations to water inaccessibility in two disadvantaged neighborhoods in Niamey, Niger. *Habitat International, 43*, 283–292.

Bossay, L. M. (2020, July 3). *Kinshasa: coronavirus, gratuité d'eau enfonce la population dans la pénurie d'eau*. Voice4Thought. https://voice4thought.org/fr/kinshasa-coronavirus-gratuite-deau-enfonce-la-population-dans-la-penurie-deau/

Bukenya, B. (2020, May 18). *What's driven Uganda's public water corporation's inadequate response to the COVID-19 crises?* Effective States and Inclusive Development Research Center. https://www.effective-states.org/whats-driven-ugandas-public-water-corporations-inadequate-response-to-the-COVID-19-crisis/

Bukuluki, P., Mwenyango, H., Katongole, S. P., Sidhva, D., & Palattiyil, G. (2020). The socio-economic and psychosocial impact of COVID-19 pandemic on urban refugees in Uganda. *Social Sciences & Humanities Open, 2*, 100045.

Bwire, G., Orach, C. G., Aceng, F. L., Arianitwe, S. E., Matseketse, D., Tumusherure, E., Makumbi, I., Muruta, A., Merrill, R. D., Debes, A., Ali, M., & Sack, D. A. (2021). Refugee settlements and cholera risks in Uganda, 2016–2019. *The American Journal of Tropical Medicine and Hygiene, 104*(4), 1225–1231. https://doi.org/10.4269/ajtmh.20-0741

Cairncross, S., Hunt, C., Boisson, S., Bostoen, K., Curtis, V., Fung, I., & Schmidt, W. (2010). Water, sanitation and hygiene for the prevention of diarrhoea. *International Journal of Epidemiology, 39*, (suppl_1), i193–i205. https://doi.org/10.1093/ije/dyq035

Chirisa, I., Mutambisi, T., Chivenge, M., Mabaso, E., Matamanda, A. R., & Ncube, R. (2020). The urban penalty of COVID-19 lockdowns across the globe: Manifestations and lessons for Anglophone sub-Saharan Africa. *GeoJournal, 87*, 1–14. https://doi.org/10.1007/s10708-020-10281-6

Collins, S., Mbullo Owuor, P., Miller, J., Boateng, G., Wekesa, P., Onono, M., & Young, S. (2019). 'I know how stressful it is to lack water!' Exploring the lived experiences of household water insecurity among pregnant and postpartum women in western Kenya. *Global Public Health, 14*(5), 649–662. https://doi.org/10.1080/17441692.2018.1521861

Cooper, R. (2020). *Water for the urban poor and COVID-19*. K4D Helpdesk Report 826. Institute of Development Studies. https://opendocs.ids.ac.uk/opendocs/handle/20.500.12413/15298

Cousins, S. (2020). 2.5 million more child marriages due to COVID-19 pandemic. *Lancet, 396*(10257), 1059. https://doi.org/10.1016/S0140-6736(20)32112-7

Danilenko, A. (2020, June 25). *Water and sanitation in Uganda: Two months when the world changed*. World Bank. https://blogs.worldbank.org/water/water-and-sanitation-uganda-two-months-when-world-changed.

Fielmua, N., Guba, B., & Mwingyine, D. (2021). Hand hygiene and safety behaviours at shopping centres in COVID-19: An observation in Wa township in Ghana. *Journal of Water, Sanitation and Hygiene for Development, 11*(3), 442–452. https://doi.org/10.2166/washdev.2021.240

Gentilini, U., Almenfi, M., & Dale, P. (2020). *Social protection and jobs responses to COVID-19: A real-time review of country measures*. Living Document Version 14. World Bank. https://openknowledge.worldbank.org/handle/10986/33635

Goujon, A., Marois, G., & Sabourin, P. (2020). Deriving Niger's demographic and education future to 2062 with stakeholders: Which results? *Population Research and Policy Review, 40*, 617–627. https://doi.org/10.1007/s11113-020-09582-y

Graham, J., Hirai, M., & Kim, S. (2016). An analysis of water collection labor among women and children in 24 Sub-Saharan African countries. *PLoS One, 11*(6), e0155981. https://doi.org/10.1371/journal.pone.0155981

Haddout, S., Priya, K., Hoguane, A., & Ljubenkov, I. (2020). Water scarcity: A big challenge to slums in Africa to fight against COVID-19. *Science & Technology Libraries, 39*, 281–288.

Hungerford, H. (2012). *Water, cities, and bodies: A relational understanding of Niamey, Niger.* (Doctoral dissertation, University of Kansas).

Ilesanmi, O., Oderinde, T., & Afolabi, A. (2020). The urban slums: Potential source of COVID-19 spikes in Africa. *Public Health in Practice, 1*, 100052. https://doi.org/10.1016/j.puhip.2020.100052

International Committee of the Red Cross (ICRC). (2020, September 15). *Democratic Republic of the Congo: Caught between COVID-19 and water shortages in Kinshasa*. ICRC. https://www.icrc.org/en/document/drcongo-caught-between-COVID-19-and-water-shortages-kinshasa

International Monetary Fund (IMF). (2021). *Policy responses to Covid-19*. https://www.imf.org/en/Topics/imf-and-covid19/Policy-Responses-to-COVID-19#N.

Karim, N., Jing, L., Lee, J., Kharel, R., Lubetkin, D., Clancy, C., et al. (2021). Lessons learned from Rwanda: Innovative strategies for prevention and containment of COVID-19. *Annals of Global Health, 87*(1). https://doi.org/10.5334/aogh.3172

Keough, S. B., & Youngstedt, S. M. (2014). The material culture of water: Transportation, storage, and consumption in Niamey, Niger. *Focus on Geography, 57*(4), 152–163.

Keough, S. B., & Youngstedt, S. M. (2019). *Water, life, and profit: Fluid economies and cultures of Niamey, Niger*. Berghahn Books.

Massinga Loembé, M., Tshangela, A., Salyer, S. J., Varma, J. K., Ogwell Ouma, A. E., & Nkengasong, J. N. (2020). COVID-19 in Africa: The spread and response. *Nature Medicine, 26*, 999–1003. https://doi.org/10.1038/s41591-020-0961-x

Mbakaya, B. C., Lee, P. H., & Lee, R. L. T. (2017). Hand hygiene intervention strategies to reduce diarrhoea and respiratory infections among schoolchildren in developing countries: A systematic review. *International Journal of Environmental Research and Public Health, 14*(4), 371–385. https://doi.org/10.3390/ijerph14040371

Mukumbang, F. C., Ambe, A. N., & Adebiyi, B. O. (2020). Unspoken inequality: How COVID-19 has exacerbated existing vulnerabilities of asylum-seekers, refugees, and undocumented migrants in South Africa. *International Journal for Equity in Health, 19*, 141. https://doi.org/10.1186/s12939-020-01259-4

Ntabadde, C. (2021). *UNICEF distributes WASH supplies to schools and health centres to contain spread of COVID-19*. UNICEF. https://www.unicef.org/uganda/stories/unicef-distributes-wash-supplies-schools-and-health-centres-contain-spread-COVID-19

Okoi, O., & Bwawa, T. (2020). How health inequality affect responses to the COVID-19 pandemic in sub-Saharan Africa. *World Development, 135*, 105067. https://doi.org/10.1016/j.worlddev.2020.105067

Olukya, G. (2020, April 1). *COVID-19: DR Congo pledges free water, electricity*. Anadolu Agency. https://www.aa.com.tr/en/africa/COVID-19-dr-congo-pledges-free-water-electricity/1787570

Olum, R., & Bongomin, F. (2020). Uganda's first 100 COVID-19 cases: Trends and lessons. *International Journal of Infectious Diseases, 96*, 517–518. https://doi.org/10.1016/j.ijid.2020.05.073

PERC (Partnership for Evidence Based Response to COVID-19). (2020, September 8). *Finding the balance: Public health and social measures in Democratic Republic of the Congo*. https://africacdc.org/download/perc-finding-the-balance-public-health-and-social-measures-in-democratic-republic-of-the-congo/

Prüss-Ustün, A., Bartram, J., Clasen, T., Colford, J., Cumming, O., Curtis, V., et al. (2014). Burden of disease from inadequate water, sanitation and hygiene in low- and middle-income settings: A retrospective analysis of data from 145 countries. *Tropical Medicine & International Health, 19*(8), 894–905. https://doi.org/10.1111/tmi.12329

Rabie, T., & Curtis, V. (2006). Handwashing and risk of respiratory infections: A quantitative systematic review. *Tropical Medicine and International Health, 11*(3), 258–267.

Saunders-Hastings, P., Crispo, J., Sikora, L., & Krewski, D. (2017). Effectiveness of personal protective measures in reducing pandemic influenza transmission: A systematic review and meta-analysis. *Epidemics, 20*, 1–20. https://doi.org/10.1016/j.epidem.2017.04.003

Sempewo, J., Mushomi, J., Tumutungire, M., Ekyalimpa, R., & Kisaakye, P. (2021). The impact of COVID-19 on households' water use in Uganda. *Water Supply, 21*(5), 2489–2504. https://doi.org/10.2166/ws.2021.044

Smiley, S. L., Agbemor, B. D., Adams, E. A., & Tutu, R. (2020). COVID-19 and water access in sub-Saharan Africa: Ghana's free water directive may not benefit water insecure households. *African Geographical Review, 39*(4), 398–404. https://doi.org/10.1080/19376812.2020.1810083

Smith, D. W., Markov, I., & Ojara, E. (2020, June 24). *Uganda's COVID-19 water sector response*. Stanford Program on Water, Health and Development. https://water.stanford.edu/news/uganda-s-COVID-19-water-sector-response

Sopi, E. (2020, March 19). *Coronavirus: la REGIDESO va augmenter sa production d'eau*. Radio Okapi. https://www.radiookapi.net/2020/03/20/actualite/sante/coronavirus-la-regideso-va-augmenter-sa-production-deau

The Independent (2020, May 21). *MPs push for free water, electricity*. The Independent. https://www.independent.co.ug/mps-push-for-free-water-electricity/

UN. (2018). *World urbanization prospects*. Department of Economic and Social Affairs. https://population.un.org/wup/

UNDP-SIWI Water Governance Facility & UNICEF. (2015). *WASH and accountability: Explaining the concept*. SIWI and UNICEF. http://www.watergovernance.org/resources/accountability-in-wash-explaining-the-concept/

UNHCR. (2021). *Where we work: Africa*. UNHCR. https://www.unhcr.org/en-us/africa.html

UNICEF & WHO. (2019). *Progress on household drinking water, sanitation and hygiene 2000–2017. Special focus on inequalities*. United Nations. https://www.who.int/water_sanitation_health/publications/jmp-report-2019/en/

USAID. (2021). *Assessing the effects of COVID-19 on access to water, sanitation, and hygiene in USAID high priority and strategy-aligned countries: Focus on small piped providers in Ghana*. https://www.globalwaters.org/resources/assets/washpals/assessing-effects-covid-19-access-water-sanitation-and-hygiene-usaid-high-priority-ghana-followon

WHO. (2017). *UN-water global analysis and assessment of sanitation and drinking-water (GLAAS) 2017 report: Financing universal water, sanitation and hygiene under the sustainable development goals*. WHO. https://apps.who.int/iris/bitstream/handle/10665/254999/9789241512190-eng.pdf

Wolf, J., Johnston, R., Freeman, M. C., Ram, P. K., Slaymaker, T., Laurenz, E., & Prüss-Ustün, A. (2019). Handwashing with soap after potential faecal contact: Global, regional and country estimates. *International Journal of Epidemiology, 48*(4), 1204–1218. https://doi.org/10.1093/ije/dyy253

World Bank. (2020, April 20). *Tackling COVID-19 (Coronavirus) with water, sanitation and hygiene in DRC*. https://www.worldbank.org/en/news/feature/2020/04/20/tackling-covid-19-coronavirus-with-water-sanitation-and-hygiene-in-drc

World Bank. (2021). *Kinshasa multisector development and urban resilience project – Kinelenda: Project appraisal document*. http://documents1.worldbank.org/curated/en/260851617415464368/pdf/Congo-Democratic-Republic-of-Kinshasa-Multisector-Development-and-Urban-Resilience-Project-Kin-Elanda.pdf

Chapter 9
Agroecology for Health: Examining the Impact of Participatory Agroecology on Health in Smallholder Farming Communities

Moses Kansanga, Daniel Kpienbaareh, Daniel Amoak, Rachel Bezner Kerr, Lizzie Shumba, Esther Lupafya, Laifolo Dakishoni, Catherine Hickey, Mangani Katundu, and Isaac Luginaah

Introduction

Amid worsening food insecurity in many parts of the world, the health of millions of people, especially women and children, has deteriorated in recent times (World Health Organization, 2018). The majority of health challenges are the result of undernourishment due to chronic food insecurity (FAO et al., 2018; Pooler et al., 2019). For instance, in 2017, the undernourished population was 821 million, implying that about one in every nine people in the world is undernourished (FAO et al., 2018). Malnutrition is responsible for about 45% of annual deaths (3.1 million) among children under 5 years, with concentration in developing countries

M. Kansanga (✉)
Department of Geography, George Washington University, Washington, DC, USA
e-mail: mkansanga@email.gwu.edu

D. Kpienbaareh
Department of Geography, Geology and the Environment, Illinois State University, Normal, IL, USA
e-mail: dlkpien@ilstu.edu

D. Amoak · C. Hickey · I. Luginaah
Department of Geography and Environment, Western University, London, ON, Canada
e-mail: damoak@uwo.ca; iluginaah@uwo.ca

R. B. Kerr
Department of Development Sociology, Cornell University, Ithaca, NY, USA

L. Shumba · E. Lupafya · L. Dakishoni
Soils, Food and Healthy Communities Organization, Ekwendeni, Malawi

M. Katundu
Chancellor College, University of Malawi, Zomba, Malawi

J. A. Braimah et al. (eds.), *Health Geography in Sub-Saharan Africa*, Global Perspectives on Health Geography, https://doi.org/10.1007/978-3-031-37565-1_9

(FAO, 2015). Approximately 151 million children under 5 years experienced stunted growth, while 50 million were faced with wasting (FAO et al., 2018). Ironically, although stunting and wasting continue to affect millions of children, the number of under-fives who are overweight reached 38 million, with 25% and 46% being from Asia and Africa, respectively (FAO et al., 2018). Similarly, adult obesity has been on the rise since 2017 (FAO et al., 2018). The rising trend of obesity in Africa is of concern due to its co-occurrence with infectious diseases such as cholera and HIV/AIDS in the context of a weak health care system (Price et al., 2018).

Although these trends are attributable to several factors, they are also a reflection of how modernity has failed to solve humanity's food needs despite tremendous advancements in modern agricultural technology (Suárez-Torres et al., 2017). The double-burden of malnutrition, as expressed in the rise in both undernutrition and overnutrition, has sparked conversations about sustainable ways of addressing global hunger. Indeed, goal 2 of the Sustainable Development Goals (SDGs) highlights the need to eradicate all forms of malnutrition and hunger. Some policymakers and researchers have advocated for agricultural intensification involving the use of synthetic inputs. However, due to poverty and inadequate access to credit, smallholder farmers are unable to adopt these input-intensive agricultural production strategies (Adenle et al., 2019; Mwangi & Kariuki, 2015). Moreover, the increased production of a few crops (e.g., corn, wheat, rice, soybean) under input-intensive agricultural systems in the context of the ongoing pursuit of a new green revolution for Africa, is progressively reshaping what smallholder communities consume (Kansanga et al., 2019a, b; Moseley, 2016). In the smallholder farming context, where households draw much of their food from the farm, diets have become less diverse (Bezner Kerr et al., 2019), with implications for health. The use of chemical inputs has also been linked to environmental degradation. Indeed, the green revolution which increased yields in some parts of the world has also caused significant damage to the physical environment, intensified the loss of agrobiodiversity and indigenous knowledge (Moseley, 2016). Moreover, the use of agrochemicals in agricultural production has also been linked to serious health problems among populations (Mnif et al., 2011). As a result of these challenges, there are calls to move toward regenerative agricultural approaches like agroecology to ensure environmentally sustainable and healthy food systems (Altieri, 2009; Altieri & Nicholls, 2017; International Panel of Experts on Sustainability Food Systems, 2016).

Agroecology is an alternative approach to agriculture that manages agroecosystems in ways that mimic the functioning of the natural ecosystem by ensuring agrobiodiversity at the farm-level and promoting energy and nutrient use efficiency (Gliessman & Engles, 2014; Kremen & Miles, 2012; Snapp, 2017). The goal is to leverage ecological synergies so that agroecosystems can replenish their own soil fertility and control pests without the use of synthetic inputs (Gliessman & Engles, 2014). Central to agroecology are agronomic practices such as mixed cropping, livestock integration, and agroforestry. Empirical research shows that farmers who adopt these practices become more productive, with positive implications on household food security, ecosystem services, and nutrition (Bezner Kerr et al., 2016; Kangmennaang et al., 2017; Kpienbaareh et al., 2020; Nyantakyi-Frimpong et al.,

2017). Agroecology also promotes farmer autonomy, participatory processes, and knowledge co-creation and sharing which can help build mutual trust among farmers with the potential to promote well-being (Dumont et al., 2016). At the broader territorial scale, agroecological approaches prioritize local, circular economies that provide employment with fair, just, decent working conditions and livelihoods (Dumont et al., 2016; Wezel et al., 2016).

Despite these benefits and the potential for agroecology to improve nutrition and, ultimately, the health of smallholder farmers, empirical studies on the health benefits of agroecology are limited, although scholars have highlighted the need for detailed analysis of its linkages with nutrition and health (Bezner Kerr et al., 2018; Kremen & Miles, 2012). The few existing studies on the linkage between agroecology and health are based on cross-sectional data (e.g. Nyantakyi-Frimpong et al., 2017). Building on the literature, we draw on longitudinal data from the Malawi Farmer-to-Farmer Agroecology Project (MAFFA) to examine the effect of agroecology on household health.

The Malawi Farmer-to-Farmer Agroecology (MAFFA) Intervention

The MAFFA project is a participatory farmer-to-farmer agroecology intervention implemented in northern and central Malawi from 2012 to 2017 following persistent widespread malnutrition and food insecurity in rural areas. Figure 9.1 provides information on the intervention sites. The project, which was funded by Global Affairs Canada, impacted about 6000 households. It harnessed local resources and used horizontal knowledge sharing about agroecology to improve food security and nutrition. At the farm-level, smallholder farmers were taught farm-level agroecology composting, manure application, crop residue integration, agroforestry, legume integration, crop rotation, and crop diversification. Cognizant of the crucial role of crop and dietary diversity in the attainment of food security, the project encouraged farmers to integrate local yellow maize, legumes, and livestock. Participating farming households were therefore resourced with a range of local seeds including yellow maize, groundnuts, sweet potato, beans, and cassava stalks/cuttings.

While farm-level agroecology practices can help diversify production and improve household diets, other relevant factors such as nutrition education and addressing social inequalities within households are crucial to improving household nutrition. Indeed, it is argued that apart from having food in the right quantities, knowledge about the right ways to combine various food constituents into a balanced diet greatly shapes the health benefits that people can draw from food (Burchi & De Muro, 2016). Furthermore, having greater gender equity within households can ensure greater access and control over food resources for women and children (Pratley, 2016). As a result, the MAFFA intervention prioritized nutrition education through a routine nutrition program dubbed ‘recipe days’. On recipe days, both men

Fig. 9.1 Showing intervention sites

and women from participating households met to discuss and experiment with new recipes and exchange knowledge on household and child nutrition. The project also organized several trainings and community-based events specifically focused on discussing gender dynamics, through the use of songs, theatre, and discussion.

As outlined earlier, the link between agroecology and health transcends improvements in household diets to the role participatory agroecology plays in fostering social interaction and the development of community ties which are relevant to both psychological and physical health. Apart from improving household nutrition, the MAFFA intervention also aimed at improving the social relations and networks of production in farming communities by promoting horizontal knowledge sharing and platforms of interaction among smallholder farmers. The rationale is that improving social context, especially fostering networks among local farmers, will in conjunction with the application of farm-level agroecology practices help improve food security, nutrition, and health in farming households. Indeed, empirical research has shown the crucial role of well-functioning social networks and trusts in local communities not just in promoting agriculture by enhancing resource sharing including farming information and inputs (Bisung et al., 2014; Lyon, 2000; Taylor & Featherstone, 2018; Van Rijn et al., 2012), but also promoting emotional health and well-being (Glanville & Story, 2018; Veenstra et al., 2005; Vincens et al., 2018). By using participatory farmer-to-farmer knowledge sharing, the MAFFA intervention also fostered social cohesion among farmers in local communities through farmer training workshops and fostering the formation of local farmer cooperatives. These programs provided opportunities for smallholder farmers to establish and nurture beneficial relationships with other farmers with such networks being potential conduits for sharing pressing concerns about agriculture and seeking both physical and emotional support (Kansanga et al., 2019a, b). Given this broad focus, the MAFFA intervention provides a unique opportunity to explore whether the participatory agroecology interventions in the Malawian context helped improve health and well-being for participating farmers.

Methods

Data

Data for this analysis are drawn from a two-wave survey data from the Malawi Farmer-to-Farmer Agroecology project and a follow-up cross-sectional survey. A stepped wedge longitudinal design was used in the MAFFA intervention. Participants were sampled from a total of 13 participating village areas in the Mzimba and Dedza districts: 10 received the intervention (5 village areas received the intervention in 2012 while 3 and 2 village areas were added sequentially in 2013 and 2014, respectively) and 3 served as controls who were never exposed to the intervention (see details of sampling on Table 9.1). Sampling into the control and intervention groups

Table 9.1 Showing study sample

Category	District	Village Area	Joined 2012	Joined 2013	Joined 2014	Total sample
Intervention	Mzimba	Chimbongondo	122			**818**
		Emtiyani	63			
		Kafulufulu	117			
		Mlimo	98			
		Kabanda		58		
		Edundu		48		
	Dedza	Mphathi	114			
		Chumachitsala		48		
		Chimoto North			105	
		Makowe			45	
Sub-total			514	154	150	
Control	Mzimba	Mtwalo	200			**400**
		Dunduzu	55			
	Dedza	Mtendere	145			
Total						**1218**

was done across separate village areas to avoid the problem of 'contamination' of intervention knowledge and materials.

In 2012 (the baseline), 2098 farming households received the intervention. Based on stepped wedge design, 350 and 440 control households received the intervention in 2013 and 2014, respectively. In a stepped wedge design where more participants are exposed to the intervention at the endline than the baseline, the effect of the intervention can be confounded with an underlying temporal trend. Conscious of this potential for confounding of the treatment effect of the intervention, the MAFFA project implemented a modified version of the stepped wedge approach such that at the endline, there were control households from the baseline who were not exposed to the intervention. This enables the estimation of the treatment effect of the intervention independent of any trend effects.

Data was collected using a longitudinal approach. In 2012, a baseline survey was conducted prior to giving out the intervention. This baseline survey included households from participating villages (n = 1218: 514 that received the intervention in 2012, 154 households that received the intervention in 2013, 150 households that received the intervention in 2014, and 400 control households that never received the intervention). In November 2017, an endline survey was conducted with the same households. In both baseline and endline surveys, the 400 control households were from 3 village areas that never received the MAFFA intervention throughout the project duration (see Table 9.1). Given the need to follow the same sample categories over time to facilitate the testing of the treatment effects of the intervention, we restricted our analytical sample to the 514 households that received the intervention in 2012 and the 400 control households that did not receive the intervention. The surveys collected information on a wide range of issues including household health, food security, nutrition, gender relations, and social networks. The survey was administered by trained enumerators who were familiar with the context and fluent in the local languages.

In July 2019 a follow-up cross-sectional survey was conducted to understand the use of the agroecological practices MAFFA farmers received training on. In total, 1033 respondents were sampled with half being from households who benefitted from the MAFFA intervention and the rest being farmers from the control group.

Measures

For the DID analysis, our dependent variable, self-rated health, was generated using four main questions that captured information on both the physical and psychosocial well-being of respondents. The first question asked participants whether they considered their physical health 'good' compared to other community members in their age group. The last three questions assessed the subjective well-being of participants based on the possession of social networks; ability to respond to crisis and having someone to confide in times of crisis; participation in community groups and having leisure. Responses to these questions are binary (Yes/No). We combined these questions to create a scale ranging from 0 to 1, where 0 implies the poorest state of health and 1 represents maximum health. The main independent variables in this analysis are membership in the participatory agroecology project and wave of survey. Regarding membership, households categorized as 'yes' received the intervention, whereas households categorized as 'no' served as the control group. Although our measures of subjective health include multiple indicators, health is a complex outcome that can be shaped by so many other socioeconomic, cultural, and political factors. Informed by the literature on well-being (Dolan & Metcalfe, 2012; Kangmennaang & Elliott, 2019), including household structure, age, and education of the primary male and primary female in the household, household food security assessed using the Household Food Insecurity Access Scale (see Knueppel et al., 2010 for a comprehensive description of this scale). Other covariates included household land ownership status, household wealth, number of crops grown, and household alcohol consumption. In the follow-up survey, participants were asked to rate their health in relation to their cohort in the community (coded 0 = poor; 1 = good).

Analysis

Our analysis is organized in two main parts. First, we used the Difference-in-Difference (DID) to estimate the treatment effect of the MAFFA intervention on self-rated health. The DID approach is a widely used approach for measuring the impact of interventions (see Grillos, 2018; Kabunga et al., 2012). The DID estimator compares the change in the outcome variable (in this case self-rated health) in the treatment group before and after the intervention to the change in the control group and estimates the average treatment effect as a linear or probit model. Our statistical approach is detailed as follows:

$$\mathrm{SRH}_{ijt} = \alpha_0 + X_{ijt}\beta_1 + X_{ijt}\beta_2 + \tau^{2012}\beta_3 + \tau^{2017}\beta_4 + \mathrm{P}_{jt}\tau^{2012}\beta_5 + \mathrm{P}_{jt}\tau^{2017} + \varepsilon_{ijt}$$

Where i is an index for an individual in household i, participating in the survey j in year t. The dependent variable SRH_{ijt} reflects the mean self-rated health score and X_{ijt} is a vector of control variables. P_j is a dummy variable, which takes the value of 1 if the household j received the intervention and 0 if in the control group. $\tau2012$ and $\tau2017$ are dummies representing the years of survey.

Although conscious efforts were made to select households with similar characteristics into both treatment and intervention groups, slight variations in baseline sample characteristics between these groups could bias the estimates. We accounted for potential selection bias by applying inversely proportional weights using the Kernel-based matching (see Kabunga et al., 2012; Kangmennaang et al., 2017). This technique matches the treatment group with a weighted average of all controls, using inversely proportional weights. Table 9.2 shows results of the balancing test for differences between participants in agroecology and non-agroecology households before and after weighting. The differences between control and intervention groups were accounted for following the weighting (see Table 9.2).

To further understand the impact of agroecology on health after the intervention, we fitted a logistic regression model using the cross-section data. Given that the health variable in the follow-up survey was binary, a logistic regression model was appropriate. The equation for our model is summarized in the following text:

$$\mathrm{Ln}\left(\frac{P}{1-P}\right) = \beta_0 + \beta_1 X_1 + \beta_2 X_2 + \cdots + \beta_k X_k$$

Table 9.2 Difference in means by participation before and after weighting

	Before weighting			After weighting		
Variable	Control	Intervention	Diff	Control	Intervention	Diff
Household size	1.164	1.038	0.12	1.062	1.038	0.024
Household structure	1.737	1.703	0.034	1.710	1.703	0.007
Age of primary male	1.369	1.409	−0.04	1.398	1.409	−0.011
Age of primary female	1.283	1.262	0.021	1.269	1.262	0.007
Education of primary male	1.287	1.189	0.098*	1.230	1.189	0.041
Education of primary female	1.226	1.184	0.042*	1.191	1.184	0.007
Household food security	0.631	0.736	−0.105**	0.758	0.736	0.022
Household wealth	1.970	1.926	0.044*	1.926	1.926	0.000
Land ownership	0.926	0.862	0.064	0.903	0.862	0.041
Farm size	0.428	0.354	0.074	0.374	0.354	0.02
Dimba/community gardening	0.530	0.412	0.118**	0.429	0.412	0.017
Alcohol consumption	1.396	1.624	−0.228	1.571	1.624	−0.053

*$p < 0.05$, **$p < 0.01$, ***$p < 0.001$

Where P is the probability of being in good health, β_0 is the constant, $\beta_1, \beta_2, \ldots, \beta_k$ are regression coefficients, and $x_1, x_2, \ldots, x_k$ are covariates. We reported findings with odds ratios (ORs), where ORs larger than 1 imply a higher likelihood of being in good health, and those smaller than 1 indicate lower odds reporting good health.

Results

Baseline sample characteristics are shown in Table 9.3. More than half of the households in both control and treatment groups were monogamous. In terms of household size, 46% of households had less than 4 members. More than two-thirds of households cultivated less 2.5 acres. In terms of wealth, 29% and 35% of MAFFA and non-MAFFA households were in the poor wealth category, respectively. More agroecology practicing households (49%) engaged in dry season gardening than non-agroecology households (43%). Also, 28% of agroecology households reported having a household member consuming alcohol on regular basis compared to 36% of non-agroecology households who reported having a member consuming alcohol regularly. More than half of both MAFFA and non-MAFFA households reported owning land.

Adjusted DID estimates of the treatment effect of agroecology on health are shown in Table 9.4. Participants in agroecology practicing households had a significant mean improvement in health compared to those from non-agroecology households. The overall treatment effect of the intervention on health was positive and statistically significant ($\beta = 0.05$, $p < 0.001$).

Findings from the logistic regression analysis are shown in Table 9.5. Consistent with the DID analysis, the results show a statistically significant association between agroecology and good health. At the bivariate level, respondents from agroecology practicing households had significantly higher odds of reporting good health (OR = 1.28, $p < 0.05$) compared to their counterparts from non-agroecology households. After controlling for demographic factors in Model 2, the significant association between agroecology and health was attenuated. However, after accounting for socioeconomic variables, including household wealth and whether the respondent's households still applied agroecological methods in farming in Model 3, the relationship between agroecology and health was statistically significant (OR = 1.501, $p < 0.05$).

Discussion and Conclusions

This chapter examined the relationship between agroecology and self-rated health using DID and logistic regression techniques. Overall, our findings demonstrate that participatory agroecology is positively associated with self-rated health:

Table 9.3 Baseline sample characteristics of households

Variable	MAFFA	Non-MAFFA	Pooled (%)
Household structure			
Monogamous	66	62	64
Polygamous	8	9	9
Separated/divorced	26	29	27
Age of primary male			
Less than 30	33	31	32
30–44	32	31	31
46–60	19	21	20
Greater than 60	16	17	17
Age of primary female			
Less than 30	37	34	36
30–44	31	33	32
46–60	18	20	19
Greater than 60	14	13	14
Education of primary male			
None	28	27	28
Primary	52	54	52
Secondary or higher	20	19	20
Education of primary female			
None	33	34	33
Primary	56	55	56
Secondary or higher	11	11	11
Household size			
Less than 4	48	43	46
4–6	34	34	34
More than 6	18	23	20
Farm size			
Less than 2.5 acres	67	70	68
2.5–5 acres	22	23	22
More than 5 acres	11	7	10
Dimba/dry season gardening			
No	51	57	54
Yes	49	43	46
Household wealth			
Poor	29	35	31
Middle	23	25	24
Rich	48	40	45
Food security			
Food secure	18	21	19
Food insecure	82	79	81
Household alcohol consumption			
No	67	58	63

(continued)

Table 9.3 (continued)

Variable	MAFFA	Non-MAFFA	Pooled (%)
Yes	33	42	37
Land ownership			
Yes	64	72	68
No	36	28	32
Total	**514**	**400**	**914**

Table 9.4 Adjusted average effects of agroecology on self-rated health

	T1 (2012)			T2 (2017)			
Outcome	Treatment	Control	Diff (BL)	Treatment	Control	Diff (FU)	Difference-in-Difference
Self-rated health	0.86	0.84	0.02	0.92	0.85	0.07***	0.05***
Robust standard errors	0.044	0.042	0.012	0.046	0,043	0.018	0.002

R^2: 0.19. *$p < 0.05$, **$p < 0.01$, ***$p < 0.001$

participants in agroecology households were more likely to report good health. The positive relationship between participatory agroecology and health in the context of smallholder agriculture can be explained by the multiple beneficial synergies inherent in agroecology practice. Agroecology is a holistic approach to agriculture that has the potential to influence health from two key dimensions: directly through improvements in household food production and nutrition, cost-savings on inputs which can be invested in health, and indirectly through the well-being benefits inherent in the social synergies participatory agroecology generates. Improved nutrition is crucial for healthy functioning (Stephenson et al., 2018). In a smallholder farming context where households draw a significant proportion of their diets from what they produced, the role of diversified production in achieving improved nutrition has been recognized (Miyamoto et al., 2019; Nyantakyi-Frimpong et al., 2017). Crop diversification is a central practice in agroecology targeted at making household diets diverse and nutritious. In the Malawian context in particular where maize monocultures have dominated smallholder agriculture in the past two decades, there is evidence of malnutrition, especially among infants mainly due to the concentration of household diets around energy-dense foods from maize (Bezner Kerr et al., 2016). While studies have demonstrated that the farm input subsidy program—the Malawian government's main policy tool for pushing the production of maize and fertilizer use—has resulted in yield increases (Kilic et al., 2014; Lunduka et al., 2013), the persistent malnutrition in this context supports the thesis that crop yield does not always result in dietary diversity and improved nutrition (Garibaldi et al., 2017). In fact, evidence from the recent Malawi Micronutrient Survey revealed that about 60% of Malawians suffer from a nutrient-related deficiency, with concentration in rural areas (Malawi National Statistical Office, 2017). Meanwhile, the country is confronted with a double burden of malnutrition given the rising obesity

Table 9.5 Logistic regression analysis of the association between agroecology and health

Variable	Model 1 Odds Ratio (SE)	Model 2 Odds Ratio (SE)	Model 3 Odds Ratio (SE)
Agroecology (Ref. No)			
Yes	1.284 (0.158)*	1.200 (0.159)	1.501 (0.269)*
Age		1.026 (0.004)***	1.035 (0.005)***
Gender			
Female		0.921 (0.133)	1.008 (0.164)
Education (Ref. No formal education)			
Primary		0.906 (0.320)	1.221 (0.461)
Secondary		0.813 (0.312)	1.315 (0.549)
Tertiary		7.453 (6.495)*	37.237 (39.728)***
Marital status (Ref. Single)			
Married		1.745 (0.606)	1.810 (0.710)
Divorced		1.685 (0.847)	1.437 (0.788)
Widowed		1.394 (0.617)	0.960 (0.484)
Household size		1.021 (0.028)	1.038 (0.031)
Gender of household head			
Female		1.657 (0.415) *	1.256 (0.381)
Household decision-making (Ref. men)			
Female head only			1.181 (0.296)
Joint decision-making			0.346 (0.059)***
Wealth (Ref. Richest)			
Richer			0.849 (0.192)
Middle			0.929 (0.191)
Poorer			0.719 (0.173)
Poorest			0.800 (0.198)
Cash savings (Ref. Yes)			
No			2.688 (0.429)***
Agroecological practices			1.145 (0.047)***
Observations		1034	1034
R^2		0.0464	0.1237

$^*p < 0.05$, $^{**}p < 0.01$, $^{***}p < 0.001$

levels (Price et al., 2018). To diversify household production and diets, the MAFFA intervention supported participating farming households with diverse seeds including pigeon pea, yellow maize, and soybean. The positive relationship from this study comes to buttress earlier studies in the Malawian context that found that crop diversity improved nutrition in HIV/AIDS-affected smallholder farming households (Nyantakyi-Frimpong et al., 2016) and child growth (Bezner Kerr et al., 2011). Moreover, given that food insecurity is a crucial social determinant of mental health (Atuoye & Luginaah, 2017; Jones, 2017; Tuthill et al., 2019), the yield and dietary improvement potential of agroecology (Kangmennaang et al., 2017; Kansanga et al., 2020) could have a positive influence on mental well-being in food insecure smallholder farming contexts.

One of the key objectives of agroecology is to foster beneficial social relations of production (Gliessman, 2016; Mdee et al., 2018). As a result, agroecology interventions like the MAFFA project prioritized improving farmer-to-farmer interaction at the local level by facilitating knowledge sharing about agriculture. The social capital inherent in this interaction can be useful in improving emotional and mental well-being. In agrarian contexts like Malawi in particular where smallholder agriculture is constantly challenged by multiple environmental stressors including droughts and floods, such social networks can provide fallback for farmers in times of crises and improve adaptive capacity (Adger, 2003; Deressa et al., 2009; Paul et al., 2016).

The findings in this study point to the crucial role agroecology can play in improving health and well-being in smallholder farming context. Indeed, our findings are consistent with other empirical studies that highlight the pivotal role of agroecology-based smallholder agriculture in improving nutrition and health (Frison et al., 2011; Kahane et al., 2013). Despite the demonstrated potential of agroecology, the lack of political will is a key barrier to its upscaling in the Malawian context. In response to the persistent food insecurity in the country, the Malawian government has for the past few decades promoted input-intensive agriculture through the Farm Input Subsidy Program that provides smallholder farmers with subsidized chemical inputs and so-called improved seeds. The persistent promotion of input intensive agriculture continues to militate against the up-take of agroecology which is both cost-effective for poor smallholder farmers and environmentally sustainable. Also, the use of a longitudinal approach adds strength to our findings given that the few studies that explored the relationship between agroecology and health have largely relied on cross-sectional data. This study provides broader lessons for approaching nutrition, health, agriculture, and environmental sustainability in a synchronized manner in the developing world, especially in the context of the ongoing pursuit of the SDGs. Given that assessment of health in the MAFFA intervention was subjective and self-reported, future research could benefit from integrating both objective and subjective measures of health.

References

Adenle, A. A., Wedig, K., & Azadi, H. (2019). Sustainable agriculture and food security in Africa: The role of innovative technologies and international organizations. *Technology in Society, 58*, 101143.

Adger, W. N. (2003). Social capital, collective action, and adaptation to climate change. *Economic Geography, 79*(4), 387–404.

Altieri, M. A. (2009). Agroecology, small farms, and food sovereignty. *Monthly Review, 61*(3), 102.

Altieri, M. A., & Nicholls, C. I. (2017). The adaptation and mitigation potential of traditional agriculture in a changing climate. *Climatic Change, 140*(1), 33–45.

Atuoye, K. N., & Luginaah, I. (2017). Food as a social determinant of mental health among household heads in the Upper West Region of Ghana. *Social Science & Medicine, 180*, 170–180.

Bezner Kerr, R., Berti, P. R., & Shumba, L. (2011). Effects of a participatory agriculture and nutrition education project on child growth in northern Malawi. *Public Health Nutrition, 14*(8), 1466–1472.

Bezner Kerr, R. B., Chilanga, E., Nyantakyi-Frimpong, H., Luginaah, I., & Lupafya, E. (2016). Integrated agriculture programs to address malnutrition in northern Malawi. *BMC Public Health, 16*(1), 1197.

Bezner Kerr, R., Rahmanian, M., Owoputi, I., & Batello, C. (2018). 6 Agroecology and nutrition: transformative possibilities and challenges. In *Sustainable diets: Linking nutrition and food systems* (p. 53). CABI.

Bezner Kerr, R., Kangmennaang, J., Dakishoni, L., Nyantakyi-Frimpong, H., Lupafya, E., Shumba, L., Msachi, R., Boateng, G. O., Snapp, S. S., & Chitaya, A. (2019). Participatory agroecological research on climate change adaptation improves smallholder farmer household food security and dietary diversity in Malawi. *Agriculture, Ecosystems & Environment, 279*, 109–121.

Bisung, E., Elliott, S. J., Schuster-Wallace, C. J., Karanja, D. M., & Bernard, A. (2014). Social capital, collective action and access to water in rural Kenya. *Social Science & Medicine, 119*, 147–154.

Burchi, F., & De Muro, P. (2016). From food availability to nutritional capabilities: Advancing food security analysis. *Food Policy, 60*, 10–19.

Deressa, T. T., Hassan, R. M., Ringler, C., Alemu, T., & Yesuf, M. (2009). Determinants of farmers' choice of adaptation methods to climate change in the Nile Basin of Ethiopia. *Global Environmental Change, 19*(2), 248–255.

Dolan, P., & Metcalfe, R. (2012). Measuring subjective wellbeing: Recommendations on measures for use by national governments. *Journal of Social Policy, 41*(2), 409–427.

Dumont, A. M., Vanloqueren, G., Stassart, P. M., & Baret, P. V. (2016). Clarifying the socioeconomic dimensions of agroecology: Between principles and practices. *Agroecology and Sustainable Food Systems, 40*(1), 24–47.

FAO. (2015). *Country fact sheet on food and agriculture policy trends*. FAO.

FAO, IFAD, UNICEF, WFP, & WHO. (2018). The state of food security and nutrition in the world 2018. In *Building climate resilience for food security and nutrition.*

Frison, E. A., Cherfas, J., & Hodgkin, T. (2011). Agricultural biodiversity is essential for a sustainable improvement in food and nutrition security. *Sustainability, 3*(1), 238–253.

Garibaldi, L. A., Gemmill-Herren, B., D'Annolfo, R., Graeub, B. E., Cunningham, S. A., & Breeze, T. D. (2017). Farming approaches for greater biodiversity, livelihoods, and food security. *Trends in Ecology & Evolution, 32*(1), 68–80.

Glanville, J. L., & Story, W. T. (2018). Social capital and self-rated health: Clarifying the role of trust. *Social Science Research, 71*(2018), 98–108.

Gliessman, S. (2016). *Transforming food systems with agroecology*. Taylor & Francis.

Gliessman, S. R., & Engles, E. W. (2014). In E. W. Engles (Ed.), *Agroecology: The ecology of sustainable food systems* (3rd ed.). CRC Press.

Grillos, T. (2018). Women's participation in environmental decision-making: Quasi-experimental evidence from northern Kenya. *World Development, 108*, 115–130.

International Panel of Experts on Sustainability Food Systems. (2016). *From uniformity to diversity: A paradigm shift from industrial agriculture to diversified agroecological systems.*

Jones, A. D. (2017). Food insecurity and mental health status: A global analysis of 149 countries. *American Journal of Preventive Medicine, 53*(2), 264–273.

Kabunga, N. S., Dubois, T., & Qaim, M. (2012). Heterogeneous information exposure and technology adoption: The case of tissue culture bananas in Kenya. *Agricultural Economics, 43*(5), 473–486.

Kahane, R., Hodgkin, T., Jaenicke, H., Hoogendoorn, C., Hermann, M., Hughes, J. d.' A., Padulosi, S., & Looney, N. (2013). Agrobiodiversity for food security, health and income. *Agronomy for Sustainable Development, 33*(4), 671–693.

Kangmennaang, J., & Elliott, S. J. (2019). 'Wellbeing is shown in our appearance, the food we eat, what we wear, and what we buy': Embodying wellbeing in Ghana. *Health & Place, 55*, 177–187.

Kangmennaang, J., Kerr, R. B., Lupafya, E., Dakishoni, L., Katundu, M., & Luginaah, I. (2017). Impact of a participatory agroecological development project on household wealth and food security in Malawi. *Food Security, 9*, 1–16.

Kansanga, M., Andersen, P., Kpienbaareh, D., Mason-Renton, S., Atuoye, K., Sano, Y., Antabe, R., & Luginaah, I. (2019a). Traditional agriculture in transition: Examining the impacts of agricultural modernization on smallholder farming in Ghana under the new Green Revolution. *International Journal of Sustainable Development & World Ecology, 26*(1), 11–24.

Kansanga, M. M., Luginaah, I., Bezner Kerr, R., Lupafya, E., & Dakishoni, L. (2019b). Beyond ecological synergies: Examining the impact of participatory agroecology on social capital in smallholder farming communities. *International Journal of Sustainable Development & World Ecology*, 1–14.

Kansanga, M. M., Kangmennaang, J., Kerr, R. B., Lupafya, E., Dakishoni, L., & Luginaah, I. (2020). Agroecology and household production diversity and dietary diversity: Evidence from a five-year agroecological intervention in rural Malawi. *Social Science & Medicine, 113550.*

Kilic, T., Whitney, E., & Winters, P. (2014). Decentralised beneficiary targeting in large-scale development programmes: Insights from the Malawi Farm Input Subsidy Programme. *Journal of African Economies, 24*(1), 26–56.

Knueppel, D., Demment, M., & Kaiser, L. (2010). Validation of the household food insecurity access scale in rural Tanzania. *Public Health Nutrition, 13*(3), 360–367.

Kpienbaareh, D., Bezner Kerr, R., Luginaah, I., Wang, J., Lupafya, E., Dakishoni, L., & Shumba, L. (2020). Spatial and ecological farmer knowledge and decision-making about ecosystem services and biodiversity. *Land, 9*(356).

Kremen, C., & Miles, A. (2012). Ecosystem services in biologically diversified versus conventional farming systems: Benefits, externalities, and trade-offs. *Ecology and Society, 17*(4).

Lunduka, R., Ricker-Gilbert, J., & Fisher, M. (2013). What are the farm-level impacts of Malawi's farm input subsidy program? A critical review. *Agricultural Economics, 44*(6), 563–579.

Lyon, F. (2000). Trust, networks and norms: The creation of social capital in agricultural economies in Ghana. *World Development, 28*(4), 663–681.

Malawi National Statistical Office. (2017). *Malawi micronutrient survey: Key indicators report 2015–16*. https://dhsprogram.com/pubs/pdf/FR319/FR319m.pdf.

Mdee, A., Wostry, A., Coulson, A., & Maro, J. (2018). A pathway to inclusive sustainable intensification in agriculture? Assessing evidence on the application of agroecology in Tanzania. *Agroecology and Sustainable Food Systems*, 1–27.

Miyamoto, K., Kawase, F., Imai, T., Sezaki, A., & Shimokata, H. (2019). Dietary diversity and healthy life expectancy—An international comparative study. *European Journal of Clinical Nutrition, 73*(3), 395.

Mnif, W., Hassine, A. I. H., Bouaziz, A., Bartegi, A., Thomas, O., & Roig, B. (2011). Effect of endocrine disruptor pesticides: A review. *International Journal of Environmental Research and Public Health, 8*(6), 2265–2303.

Moseley, W. G. (2016). The new green revolution for Africa: A political ecology critique. *The Brown Journal of World Affairs, 23*, 177.

Mwangi, M., & Kariuki, S. (2015). Factors determining adoption of new agricultural technology by smallholder farmers in developing countries. *Journal of Economics and Sustainable Development, 6*(5).

Nyantakyi-Frimpong, H., Mambulu, F. N., Kerr, R. B., Luginaah, I., & Lupafya, E. (2016). Agroecology and sustainable food systems: Participatory research to improve food security among HIV-affected households in northern Malawi. *Social Science & Medicine, 164*, 89–99.

Nyantakyi-Frimpong, H., Kangmennaang, J., Kerr, R. B., Luginaah, I., Dakishoni, L., Lupafya, E., Shumba, L., & Katundu, M. (2017). Agroecology and healthy food systems in semi-humid tropical Africa: Participatory research with vulnerable farming households in Malawi. *Acta Tropica, 175*, 42–49.

Paul, C. J., Weinthal, E. S., Bellemare, M. F., & Jeuland, M. A. (2016). Social capital, trust, and adaptation to climate change: Evidence from rural Ethiopia. *Global Environmental Change, 36*, 124–138.

Pooler, J. A., Hartline-Grafton, H., DeBor, M., Sudore, R. L., & Seligman, H. K. (2019). Food insecurity: A key social determinant of health for older adults. *Journal of the American Geriatrics Society, 67*(3), 421–424.

Pratley, P. (2016). Associations between quantitative measures of women's empowerment and access to care and health status for mothers and their children: A systematic review of evidence from the developing world. *Social Science & Medicine, 169*, 119–131.

Price, A. J., Crampin, A. C., Amberbir, A., Kayuni-Chihana, N., Musicha, C., Tafatatha, T., Branson, K., Lawlor, D. A., Mwaiyeghele, E., & Nkhwazi, L. (2018). Prevalence of obesity, hypertension, and diabetes, and cascade of care in sub-Saharan Africa: A cross-sectional, population-based study in rural and urban Malawi. *The Lancet Diabetes & Endocrinology, 6*(3), 208–222.

Snapp, S. (2017). Agroecology: Principles and practice. In *Agricultural systems* (pp. 33–72). Elsevier.

Stephenson, J., Heslehurst, N., Hall, J., Schoenaker, D. A. J. M., Hutchinson, J., Cade, J. E., Poston, L., Barrett, G., Crozier, S. R., & Barker, M. (2018). Before the beginning: Nutrition and lifestyle in the preconception period and its importance for future health. *The Lancet, 391*(10132), 1830–1841.

Suárez-Torres, J., Suárez-López, J. R., López-Paredes, D., Morocho, H., Cachiguango-Cachiguango, L. E., & Dellai, W. (2017). Agroecology and health: Lessons from indigenous populations. *Current Environmental Health Reports, 4*(2), 244–251.

Taylor, M. R., & Featherstone, A. M. (2018). The value of social capital in farmland leasing relationships. *Agricultural Finance Review, 78*(4), 489–496.

Tuthill, E. L., Sheira, L. A., Palar, K., Frongillo, E. A., Wilson, T. E., Adedimeji, A., Merenstein, D., Cohen, M. H., Wentz, E. L., & Adimora, A. A. (2019). Persistent food insecurity is associated with adverse mental health among women living with or at risk of HIV in the United States. *The Journal of Nutrition, 149*(2), 240–248.

Van Rijn, F., Bulte, E., & Adekunle, A. (2012). Social capital and agricultural innovation in Sub-Saharan Africa. *Agricultural Systems, 108*, 112–122.

Veenstra, G., Luginaah, I., Wakefield, S., Birch, S., Eyles, J., & Elliott, S. (2005). Who you know, where you live: Social capital, neighbourhood and health. *Social Science & Medicine, 60*(12), 2799–2818.

Vincens, N., Emmelin, M., & Stafström, M. (2018). Social capital, income inequality and the social gradient in self-rated health in Latin America: A fixed effects analysis. *Social Science & Medicine, 196*, 115–122.

Wezel, A., Brives, H., Casagrande, M., Clement, C., Dufour, A., & Vandenbroucke, P. (2016). Agroecology territories: Places for sustainable agricultural and food systems and biodiversity conservation. *Agroecology and Sustainable Food Systems, 40*(2), 132–144.

WHO. (2018). *World health statistics 2018: monitoring health for the SDGs, sustainable development goals*.

Chapter 10
Cities that Work for All: Putting Gender on the Urban Health and Development Agenda in Sub-Saharan Africa

Eunice Annan-Aggrey, Senanu Kwasi Kutor, Elmond Bandauko, and Godwin Arku

Introduction

Gender and women's socio-economic empowerment are core development themes at the global level. The inequities women experience underscores the importance of gender responsive development. For instance, women continue to be at the bottom of the economic ladder with 50% less wealth than men (Oxfam, 2021). Nonetheless, women provide 12.5 billion hours of unpaid care work every day, which is three times more than men do (Oxfam, 2020). Moreover, women make up over half of the global population, yet they continue to suffer systematic gender discrimination in cities (UN-Habitat, 2020). Currently, cities are home to more than half of the world's population, and this percentage is expected to increase to 60% by 2030 (UN-DESA, 2016). Hence, the salience of gender responsive urban development for women's health cannot be overemphasized. This is because gender responsive development contributes to promoting gender equality and securing sustainable livelihoods for a significant proportion of global citizens.

The SDGs reflect a commitment to respect and protect human rights and acknowledge the important links between inequality, marginalization, and poverty (Esquivel & Sweetman, 2016). The SDGs underscore the central role of gender equality by identifying it as a cross-cutting theme with relevance for all the 17 interconnected goals.

Additionally, Goal 11 of the SDGs seeks to make cities and human settlements inclusive, safe, and resilient. This also involves enhancing access to adequate, safe, and affordable housing and basic services for all urban residents, including women (UN, 2015a, b). SDG 11 also aims to upgrade urban slums, where women are

E. Annan-Aggrey (✉) · S. K. Kutor · E. Bandauko · G. Arku
Department of Geography and Environment, Western University, London, ON, Canada
e-mail: eannanag@uwo.ca; skutor@uwo.ca; ebandauk@uwo.ca; garku@uwo.ca

J. A. Braimah et al. (eds.), *Health Geography in Sub-Saharan Africa*, Global Perspectives on Health Geography, https://doi.org/10.1007/978-3-031-37565-1_10

overrepresented. Specifically, target 11.7 aspires to provide "universal access to safe, inclusive and accessible, green and public spaces, in particular for women and children, older persons and persons with disabilities by the year 2030" (UN, 2015a, b, p. 26). Furthermore, the New Urban Agenda (NUA) also commits to including women in sustainable development by ensuring that the physical and social infrastructure of cities, including water and sanitation, and housing are responsive to the needs, rights, and safety of urban women and girls (UN-Habitat, 2020). The foregoing discussion emphasizes that sound gender equality systems are imperative to promoting human rights, health, and dignity, as well as to achieving prosperity and inclusive growth in cities and urban areas (UN Women, 2018).

Despite the existence of global policy frameworks on gender equality and inclusive cities, women, and especially those in developing countries, continue to experience socio-economic marginalization. In most cities in sub-Saharan Africa (SSA), the struggle to make urban development responsive to the needs of women living in poverty continues (Brouder & Sweetman, 2015). SSA represents the region with the highest percentage of urban dwellers (56%) living in underserved slum areas (UN, 2017). In some SSA countries such as Ghana and Lesotho, there are more women than men (120:100) between the ages of 15 and 49 living in slum conditions (Azcona et al., 2020) (See Table 10.1). This suggests that urban sex ratios are getting more feminized in some countries in SSA (IOM, 2015). While ratios of men and women in urban areas are at par, the higher number of women living in slums is disturbing. This is because inadequate housing and access to services tend to have a greater impact on women, who spend more time in the home and its immediate environment than men (Chant, 2003). Additionally, in such contexts, women face challenges and systemic barriers that result in unequal access to and control over economic resources that deepen their socio-economic marginalization.

The urbanization process often creates new development and health opportunities as well as challenges. On the positive side, ever-expanding cities present opportunities to ameliorate gender inequalities by creating more employment opportunities for women. At the same time, challenges such as the uneven distribution of services are exacerbating existing gender inequalities (Meleis, 2011). Additionally, many women migrants to urban areas move away from their families and local support system, becoming isolated and vulnerable in their new environments (Awumbila, 2015). Existing gender inequalities and gender roles

Table 10.1 Slum populations in Ghana and Lesotho

Characteristic/Country	Ghana	Lesotho
Total population (2014)	26,442,000	2,098,000
Population living in urban areas	53% (14,118,000)	26.8% (562,000)
Urbanization rate	3.4%	3.05%
Urban population living in slums	37.9%	50.8%
Population of urban slum dwellers	5,349,000	285,496

Source: UN-Habitat Slum Almanac 2015/2016

make women more susceptible to urban development challenges, including the distribution of housing, planning, and the delivery of services such as water and sanitation. According to UN-Habitat (2020, p. 149), "gender sensitive urban policies can benefit women in diverse ways, including, improving access to services and infrastructure, and removing gendered socio-cultural restrictions." Therefore, gender-sensitive urban policies ensure that women reap the same benefits associated with urban living as men.

This chapter discusses the significance of mainstreaming gender in urban planning, housing, and service delivery (water and sanitation), using examples from SSA. It also examines the main structural barriers working against gender responsive urban development and identifies key policy pointers to achieve gender equality in cities. The chapter starts with an analysis of the linkages between urban planning and gender equality. The next section focuses on gender and urban housing, followed by a discussion on the gender dimensions of water and sanitation service delivery. The chapter concludes with policy recommendations that can be adopted by urban policymakers for gender responsive urban development.

Gender, Cities, and Urban Planning

Women's empowerment and gender equality are prominent issues in city planning and architecture (Sánchez de Madariaga & Neuman, 2016). Horelli (2017, p. 1780) argues that "the gender perspective is crucial for planning not only due to its emphasis on the issues of equity and equality for having a voice in decision-making, but also because of its focus on the social complexities of everyday life." Cities require gender-transformative urban planning to ensure their infrastructure can support the needs and aspirations of all citizens, regardless of gender and other aspects of their identity (Escalante & Valdivia, 2015). Beall (1996, p. 9) argues that in order to develop participatory processes that incorporate both men and women at all levels of urban development, there is the need for a new system of urban cooperation. In the planning and design of urban space, participatory planning processes must factor in women's needs according to their realities (Kabeer, 2005). Our view of urban planning from a gender perspective is that women are experts about places where they live; they have a richer lived experience about neighborhoods and cities.

However, the practice of urban planning has traditionally been 'gender blind,' creating built environments that do not adequately address the different needs between men and women (Malaza, 2009). City planners have for a long time assumed the working male to be the primary user of cities. Thus, many city spaces have been designed in a 'gender-blind' manner (Feinstein & Servon, 2005; Yon & Nadimpalli, 2017). Women's priorities have often been ignored in the design of human settlements, the location of housing, and the provision of urban services (Beall, 1996). The impediments to gender equality manifest themselves in different forms, such as inadequate infrastructure, violence against women in public spaces, and inadequate representation in urban governance structures. These bottlenecks negatively impact women's participation in urban economies, affect their health, and constrain their development capacities (Chant, 2013; Razavi & Turquet, 2016; UN-Habitat, 2011). For example, a study by ActionAid International (2013) discovered that in the Democratic Republic of Congo (DRC), there is a lack of legislation to protect women from violence in public spaces, and women are rarely involved in urban governance and planning. In addition, poor urban planning which results in the unavailability of water and sanitation services imposes a disproportionate psychosocial burden on women (Bisung & Elliott, 2017). This is because women have more care functions that require the use of water, such as cooking and cleaning the home; thus, the absence or inadequate supply of water negatively affects the health of women causing distress and anxiety.

Another area in city planning that has gendered implications is transportation. Women have different mobility needs and travel patterns. However, it is important to note that their mobility is challenged by physical, economic, cultural, and psychological constraints and transportation policies that disregard women's needs (Sánchez de Madariaga & Neuman, 2016). Women's safety and security concerns in cities is another factor that is closely related to issues of exclusion and marginalization. In developing countries, there are increased cases of gender-based violence in public spaces and transportation (Moser & McIlwaine, 2004). By virtue of their socio-economic status, women feel more unsafe and insecure than men. In cities such as Lagos in Nigeria, women engage in night-time commercial activities from food vending to providing hair care services. The risk of being attacked is very high and these women contend with challenges commuting around the city. Lagos was voted the eighth most dangerous city for women in the world (ActionAid International, 2013). This sense of insecurity negatively affects women's everyday life, restricting their personal freedom and access to urban public spaces associated with employment, health, education, and political and recreational facilities. These challenges have a huge impact on women from low-income urban communities, especially in developing countries (Moser, 2012). A survey conducted by Cities Alliance in Greater Banjul reveals that women and girls usually experience sexual harassment in public transport (Cities Alliance, 2020). Moreover, women and girls continue to be marginalized in urban governance processes (IWPR, 2015). The growth of the informal economy and slum settlements affect women disproportionately, not only because they are generally poorer than men, but also because they lack decision-making opportunities and experience greater difficulty accessing resources and services tailored to their needs (UN Habitat, 2015a). These barriers contribute to the range of obstacles women face in participating in social, economic, and political processes.

The traditional zoning systems based on separate land-use have led to fragmented urban environments, where residences, workplaces, shopping, and leisure districts are in separate spaces linked by extensive road and transportation systems. These practices have been criticized for isolating women at home, making access to services and facilities inconvenient. A study conducted in Durban's low-income settlements reveal that female-headed households were more seriously affected by planners' decisions on the number and location of healthcare facilities provided in low-income areas (Venter et al., 2007). All these issues trigger questions such as: whose city is it anyway? Do women have the right to the city? The major structural barrier to achieving any notion of rights to the city is asymmetrical power relations. Other barriers that impede the right to the city include violence (Whitzman et al., 2013), and other structural barriers (UN-Habitat, 2015a) which impact citizens' ability to participate in city life fully and freely.

The 'gender blindness' of current planning models and approaches (Fenster, 2005) is also a key barrier and demonstrates the extent to which women's rights to full participation in urban life are denied. For example, the right to appropriate urban public space in cities is inherently gendered. The right to the city, from this standpoint, fails to address multiple disadvantages and multiple forms of discrimination in everyday life. Meth (2010) argues that women living in Cator Minor and Warwick Junction in Durban, South Africa, continue to endure precarious living conditions given that they have had to contend with multiple evictions; clearly, their 'right to the city' is not guaranteed. Similarly, migrant women in Johannesburg's inner city face structural barriers that inhibit them from fully participating in urban social and economic life (Kihato, 2011). In Dakar, Senegal, a road enlargement project resulted in differential outcomes for individual street traders. Because of their social class position, women's vending spaces were eliminated, and they moved to less lucrative suburban markets (Bass, 2000). Similarly, in 2017, the Parkyard market in central Kampala was deemed "illegal" and demolished. A private company took over to redevelop the area into a "modern" shopping complex. This demolition disrupted thousands of livelihoods, with women, the elderly, and the poorest struggling to bounce back (Adama & Lindell, 2020). The Parkyard market demolition shows that when neo-liberal agendas of urban modernity prevail, women and other groups of the urban poor are relegated to the margins of the urban economy and their 'right to the city' thwarted.

Gender and Urban Housing

Housing is one of the most basic human needs, hence, access to adequate housing is recognized as a fundamental human right essential for human security, nutrition, and health (UN-Habitat, 2016). Despite this provision, over a billion people are inadequately housed, with several millions living in overcrowded slums and informal settlements. The global housing challenge is most acute in SSA, where the average annual urban population growth rate of 4.1% outstrips the global average urban population growth of 2.0% (Saghir & Santoro, 2018; World Bank, 2019). The rapid urban growth is accompanied by a spiraling demand for land and housing units. The provision of housing units and related infrastructure in cities by national and local governments, therefore, struggles to keep pace with the demand. Hence, there is a significant housing gap in many SSA cities, and the housing sector is characterized by inadequate housing units, overcrowding, and housing decay (Arku,

2009; Boamah, 2014; Tusting et al., 2019). For instance, Kenya's housing shortage is estimated at 2 million units, while Nigeria has a deficit of 17 million housing units (IFC, 2021). Similarly, the housing challenge in urban Ghana includes both a backlog in the provision of formal housing and the resultant spread of informal settlements (Arku et al., 2012; Boamah, 2014; Karley, 2008). The continued upward trajectory of SSA's urban population however suggests that the housing deficit is expected to grow even wider. Population projections indicate that urban dwellers will continue to outnumber rural residents, with an additional 560 million people estimated to move into SSA cities by 2040 (Saghir & Santoro, 2018).

The increasing housing deficit in SSA cities has a significant impact on the livelihood of poor households, particularly women. This is because affordability is a key component of the right to adequate housing. An assessment of land and housing in SSA by UN-Habitat (2011) shows that with the exception of a few countries, including South Africa, Mali, and Ethiopia, there is little evidence of the provision of affordable large-scale land and housing in SSA. In many cases, affordable housing programs either do not exist or do not produce enough units to meet demand. The challenge of limited affordable housing options results in poor urban residents, especially women, resorting to informal settlements and slum dwellings. Consequently, in East and West Africa, at least half of the urban population lives in slum areas and informal settlements, while in Central Africa, 3 out of 5 urban persons reside in slums (UN-Habitat et al., 2020). Recent surveys suggest that women are overrepresented among the population with inadequate housing in SSA cities (Azcona et al., 2020; COHRE, 2008; World Bank, 2015a, b). For example, in Kenya, there are 116 women for every 100 men living in slum conditions; while in South Africa, there are 119 women for every 100 men residing in informal settlements and slum areas (Azcona et al., 2020). Further analysis suggests that while there may be comparable numbers of men and women in urban areas, informal settlements have more women than men. The situation in Ghana shows that there are 30 more women for every 100 men living in slum areas, for the 15 to 49 age group (Azcona et al., 2020). Overall, more than 50% of urban women in SSA cities live in inadequate housing facilities with insufficient living areas. Additionally, many lack access to basic services, including clean water and improved sanitation facilities (UN Women, 2018). Hence, a significant proportion of women residing in slums in SSA cities do not enjoy the fundamental right to adequate housing (OHCHR, 2012). Furthermore, slum residents, including women, are being left behind across a range of well-being outcomes, such as access to health services and education.

Housing scholars, including Zack et al. (2003), argue that the right to adequate housing goes beyond the provision of shelter to "encompass social concepts including privacy, dignity and safety and status" (p. 45, 50). The characteristics of informal settlements in slum areas however do not meet the physical standards of adequate housing (Table 10.2); let alone satisfy the psychological need for a decent place to call 'home' (Zack et al., 2003). Thus, it can be argued that urban slum dwellers in SSA cities who do not have access to shelter facilities that offer the full functions of a 'home' are 'homeless.'

Table 10.2 Five key shelter deprivations as measures of slums

Key indicators	Definition
Structural quality/ durability of dwellings	A house is considered 'durable' if it is built in a non-hazardous location and has a structure that is permanent and adequate enough to protect its inhabitants from the extremes of climatic conditions such as rain, heat, cold, and humidity
Sufficient living area	A house is considered to provide a sufficient living area for the household members if not more than three people share the same room
Access to improved water	A household is considered to have access to improved water supply if it has a sufficient amount of water for family use, at an affordable price, available to household members without being subject to extreme effort, especially to women and children
Access to improved sanitation	A household is considered to have access to 'improved' sanitation if it has a human excreta disposal system, either in the form of a private toilet or a public toilet shared by a maximum of two households. In urban areas, access to improved sanitation is defined by direct connection to a public, piped sewer; direct connection to a septic system; or access to pour-flush latrines or ventilated improved pit latrines, allowing for acceptable local technologies
Security of tenure	Secure tenure is the right of all individuals and groups to effective protection by the state against forced evictions. International law defines forced eviction as *"the permanent or temporary removal against their will of individuals, families and/or communities from the homes and/or land which they occupy, without the provision of, and access to appropriate forms of legal or other protection."*

Source: UN-HABITAT (2011, p. 19)

Slum dwellers do not have permanent structures because of the lack of access to tenure over the lands they occupy (UN Habitat, 2011). For women and girls especially, the temporary nature of slum dwellings does not offer adequate security, but rather makes them susceptible to gender-based violence and other forms of abuse which exacerbate gender inequality (Goebel, 2015). Moreover, the informal structures used by slum residents are more prone to adverse weather conditions such as flooding and extreme heat (Ajibade et al., 2013). Thus, women who spend more time at home due to their caregiving roles are more exposed to the health and hygiene risks posed by these makeshift accommodations (COHRE, 2008; Venter & Marais, 2006). For example, a study in flood-prone areas in Old Fadama, a slum in Accra, Ghana, confirmed that women and children were affected more directly by the impacts of flooding as they spent more time cleaning up the flood debris as well as taking care of household members affected by the flood (Adams & Nyantakyi-Frimpong, 2020).

Gender inequality issues, including the disproportionate representation of women among the poor, contribute to many women's inability to enjoy adequate housing in SSA cities. World Bank (2020) data demonstrates that many countries in SSA have higher numbers of poor women than men. For instance, for the bottom 40% of the income spectrum, Botswana has 42% women compared to 38% men. For the same income quintile (bottom 40%), Burundi, Cote d'Ivoire, Namibia, Rwanda, and Sierra Leone all have 41% women compared to 39% men. McFerson (2010) identifies restricted property rights of women as one of the major causes of women's poverty in SSA. Lack of access and control over land and other economic assets makes more women prone to experience poverty. In many parts of SSA, women's access to affordable land and housing is dependent on their relationships with men as wives or daughters (World Bank, 2015a, b). Consequently, when these relationships are severed or women are widowed, many women and girls experience discrimination in inheritance; this impacts their continued ability to enjoy the right to adequate housing. Such discriminatory practices are often embedded in customary practices which fail to recognize women as legitimate successors or recognize women as having equal rights as men to inheritance (OHCHR, 2012). Inheritance provisions that do not recognize women as legitimate successors deprive women of their property (Awumbila, 2015). This situation presents women with limited housing options and other land-based resources. Losing access to land and housing increases women's vulnerability and exacerbates their poverty (Azcona et al., 2020). Furthermore, inheritance issues have been identified among the major push factors contributing to the increasing number of female migrants to urban areas in SSA (COHRE, 2008). For instance, a survey among slum residents in two major cities in Ghana, Accra, and Kumasi demonstrated that many women who experienced inheritance conflicts and lost control over the property of their husbands or fathers in northern Ghana migrated to southern cities such as Accra and Kumasi, to start life anew (COHRE, 2008, p. 13). Many of these women were however unable to afford the rental charges for formal housing units and so ended up in slum settlements. Research in Kenya (COHRE, 2008) and in South Africa (Goebel, 2015) also confirm that more women are migrating to urban areas, hence the demand for informal urban housing is being catapulted by rural-urban migration. Further analysis suggests that harmful gender-based cultural practices, including child early and forced marriage, serve as triggers to stimulate women's migration to urban areas. In other cases, stigmatization by community members precipitates women's migration. For example, some women living in slums in Nairobi, Kenya, indicated that they lost their husbands to HIV/AIDS and decided to migrate in order to evade the condemning stares of their neighbors who presumed that they had been infected and so had limited time to live (COHRE, 2008).

Ironically, while many women migrants attempt to escape harmful practices in their native communities by moving to urban areas, their new homes, however, often exacerbate their vulnerability and expose them to further marginalization. The vision to seek economic opportunities to fend for themselves is often unmet because other intersectional characteristics contribute to marginalizing women and reducing the decent housing options available to them in SSA cities. Residing in inadequate housing facilities in turn expose women to vulnerabilities that deepen gender inequality, such as weak security systems which makes them liable to sexual abuse (Goebel, 2015). The optimism of female migrants that cities will present better economic opportunities for them is often short-lived because many migrants move from an agrarian rural economy to urban areas and do not possess skills suited for work in cities (COHRE, 2008). They, therefore, take up menial informal jobs, mainly which pay meagre wages that are not sufficient to rent housing units (Awumbila & Ardayfio-Schandorf, 2008). Thus, they resort to slum areas which are largely underserved with adequate basic service delivery, including water and sanitation (COHRE, 2008). In Accra, Ghana, for example, young women migrants who serve as head porters (Kayayei) in markets organize themselves into groups of 10–15 to rent kiosks and shacks (Tufuor, 2009). These dormitory accommodations often do not have access to water and toilet facilities, and this exposes these women to gender-based violence as some men take advantage and attack and sexually abuse them during their late night and dawn treks to public places of convenience (COHRE, 2008). Goebel (2015) also highlights a similar situation in informal settlements in urban South Africa where the open/unfenced nature of settlements exposed women to theft and assault. The case of female headed households is more precarious because some men take advantage of the absence of adult males in the households to abuse women and girls. Hence, women escaping gender inequalities experience further vulnerability due to their inability to access adequate housing. Among the major deprivations experienced by women living in inadequate housing facilities is a lack of access to water and sanitation. The following sections, therefore, discuss the impact of water and sanitation services delivery on women in urban areas.

Gender and Urban Service Delivery: A Focus on Water and Sanitation

The water and sanitation sector plays an important role in enhancing the health and well-being of urban communities. According to the World Bank (2006), satisfactory targeted social services can improve women's health and security and that of their families and better allow them to participate in economic, social, and political activities. Many municipalities in SSA, however, face serious constraints in delivering adequate urban services, such as water and sanitation. In most cases, it is women who bear the consequences of inadequate urban service provision due to their socially assigned roles, including water collection for cooking and cleaning (Venter et al., 2007). SDG 6 (target 6.2) sets an ambitious target for all countries to attain by 2030: "achieve adequate and equitable sanitation and hygiene for all and end open defecation, paying special attention to the needs of women and girls and those in vulnerable situations" (UN 2015a, b: 20). However, several countries in SSA face challenges in providing requisite and improved sanitation for citizens (UNICEF, 2017). For instance, the WHO/UNICEF 2017 update on progress on drinking water, sanitation and hygiene indicates that in 2015, only 19% of the urban population in Ghana had access to basic sanitation services (WHO & UNICEF, 2017). Similarly, Congo recorded only 20% coverage, while Ethiopia recorded 18% sanitation coverage for the urban population in 2015 (WHO/ UNICEF, 2017). In the context of urban Ghana, a study by Peprah et al. (2015), in four neighborhoods in Accra (Alajo, Bukom, Old Fadama, and Shiabu), demonstrated a significant variation in household sanitation coverage between the neighborhoods surveyed. For instance, Old Fadama which is a slum area with informal settlements indicated lower household sanitation coverage with most households depending on shared public toilet facilities. Women are the most affected by lack of sanitation in informal settlements (COHRE, 2008). For example, 30% of female respondents in a study in the Mathare Valley Informal Settlement in Nairobi, Kenya, reported at least one case of diarrhea in the month preceding the study (Corburn & Hilderbrand, 2015).

The extant literature demonstrates that the lack of access to adequate sanitation has grievous implications for the psychosocial well-being of individuals and households. Women in general endure a disproportionate burden of stress emanating from the lack of adequate sanitation compared to men in many situations (Bisung & Elliott, 2017). A study in a school setting in South Africa revealed that most girls perceived the use of public toilets as dangerous, especially given their experiences of harassment by their male teachers and their colleagues. Hence, the girls used unhygienic alternatives and did not patronize the shared facilities (Abraham et al., 2006). Furthermore, in Usoma, a lakeshore community in Kenya et al. (2016) observe that women showed deep-seated feelings of embarrassment, anxiety, and frustration due to their daily lived experiences associated with lack of adequate and improved sanitation. These feelings in relation to their well-being are summarized by Marmot (2014) as the psychological experience of inequality which has deep impacts on the mind of individuals.

The role of social and physical barriers in accessing public services cannot be overlooked. The combined effects of social and physical barriers restrain women from accessing urban services such as water (UN-Women, 2020). This observation is critical because safe water and sanitation are preconditions to good health, development, and human growth (Agbadi, Darkwah & Kenney, 2019; Gomez, Perdiguerro & Sanz, 2019; Fisher et al., 2015; WHO/UNICEF, 2013; WHO, 2011). Gomez et al. (2019) note that globally, 844 million people still lack access to basic drinking water, despite considerable progress in the delivery of improved water services between the years 2000 and 2015. Thus, rivalry and exclusion over water are rampant in urban areas due to greater demand, resulting from population growth (Muller, 2008). Global water and sanitation practitioners, therefore, acknowledge the need to include gender-based perspectives in policy planning because women and girls are the primary providers and managers of water at the household level (WSP, 2010). A study to examine water transportation in 15 countries in SSA showed that women were the primary collectors of water in over half of the households sampled (Sorenson et al., 2011). Consequently, improved water services benefit women and girls the most because when water systems are non-functional, women are compelled to trek long distances, over many hours, in search of water (UN-Women, 2013). According to UNDP (2006), 40 billion hours a year are collectively spent fetching water in SSA, equivalent to a year's labor for the entire workforce of France. Additionally, scholarly work that explored the unequal burden of water privatization and women's human rights in Tanzania revealed that women were faced with the complex challenges of living in poverty and lacking access to water (Brown, 2010). This finding highlights the need to include women's views from the design phase of water service projects through to the implementation stages (Brown, 2010).

Conclusion: Towards Cities that Work for All

In this chapter, we have demonstrated the salience of gender equality for the achievement of sustainable development targets. In many cities in SSA, low-income earners, particularly women, reside in informal settlements characterized by erratic service delivery. Given that women experience cities differently from men, the one size fits all approach to urban development does not sufficiently address existing social inequalities. Therefore, in order to address the structural barriers that exist in the context of urban planning, housing, and service delivery in cities in SSA, the following section suggests strategies that city authorities and other related duty bearers can adopt toward a more gender-equitable urban development.

First, urban development organizations working on gender issues should explore opportunities to engage with grassroots women directly. These strategic partnerships are critical in ensuring that women's voices are clearly articulated, and their needs reflected in the planning, governance, and management of urban areas. For example, the Slum Women's Initiative for Development (SWID) in Uganda successfully negotiated with local authorities in the Jinja area to commit to the Resilient Cities Campaign led by the United Nations Office for Disaster Risk Reduction (UNISDR). Engagement with this initiative led to infrastructure improvements in the area's informal settlements. Additionally, the SWID mobilized community-based organizations, local leaders, and grassroots women to amplify community concerns in municipal planning and governance (World Bank, 2015a).

Second, municipalities in SSA should endeavor to build inclusive cities through the design of urban development policies, plans, and regulations that are sensitive to the needs of all citizens, including women. These policies and plans could specifically endorse the interventions needed to create inclusive urban spaces. Examples of these interventions include: (i) adoption of gender-sensitive urban governance models, to create spaces for women to make their voices heard in the planning of cities and urban spaces, and (ii) working toward strengthening land tenure security for women, especially those living in poor urban settlements like slums. For example, a Cities Alliance initiative in informal settlements in the Greater Monrovia Area in Liberia incorporated women's ideas and needs into the formulation of a City Development Strategy. The participatory approach employed enhanced diversity and strengthened the voices of women who had hitherto experienced socio-cultural discrimination (World Bank, 2015a). Additionally, water kiosks constructed in the area were placed under the care of women, contributing to transformative gender dynamics in the area by empowering women to take up leadership roles in the water committees. In Greater Banjul, Gambia, Cities Alliance developed the Women's Engagement City (WEC) profile, a tool for urban stakeholders, women, and policy makers to guide the collection and analysis of data on how women and girls experience the city. Cities Alliance implemented a variety of participatory processes to ensure that women were able to share their experiences of the city and voice their concerns and wishes for the new city plan (Cities Alliance, 2020).

Third, urban authorities must adequately implement strategies that strengthen women's 'right to the city,' thereby empowering them. This includes helping them to improve their levels of civic participation that provide them with a greater feeling of security and safety. A rights-based approach to women's safety means that women should have the right to access land and housing, be protected from homelessness and forced eviction, have the right to freely and safely move within cities, and receive access to basic services such as water, sanitation, and hygiene.

Fourth, municipalities must adopt and institutionalize gender-responsive budgeting as a tool to sensitize elected officials about the specific needs of women. Involving women in municipal budgeting processes increases their capacity to participate in public decision-making processes. Women's active participation in municipal budgeting processes will ensure that their needs and priorities are captured in the planning and delivery of urban services. For example, the Zimbabwe Women Resource Center and Network (ZWRCN) has been working with municipalities in Zimbabwe (Bulawayo, Kwekwe, Kadoma) to build the capacity of both officials and councilors in mainstreaming gender in budget policymaking (ZWRCN, 2015). The capacity building activities have strengthened the analytical capacity of citizens and other stakeholders such as community-based organizations to engage local governments and demand accountability for gender equality commitments.

Fifth, it is important for urban governments to adopt gender-sensitive governance, which involves both the substantive representation of women in urban decision-making and enhanced awareness and understanding of gender-specific needs within the governance structure. This process involves elevating women's voices and participation in decision-making and their agency to effect urban policy change. Gender-sensitive governance identifies gaps in policy and service provision that disproportionately affect women, acknowledging their unique contribution to urban settings in the formation of policy responses. It involves the meaningful interaction of government representatives with grassroots women's movements and civil society groups that actively advocate for gender equality. There is also the need for effective and well-coordinated gender-responsive programming in the Water, Sanitation, and Hygiene (WASH) sector in SSA. This can contribute to gender equality and equity in accessing urban WASH services. According to UNICEF (2017), gender mainstreaming in the WASH sector improves gender and WASH results through effective administration and functioning of water supply, improved cleanliness in schools, reduced open defecation, increased women's contribution and voice in decision-making, and increased school attendance for girls. For example, a World Bank evaluation of 122 water projects in SSA found that the effectiveness of a project was six to seven times higher when women were involved compared to when they were not (Ngum, 2018).

These policy recommendations will go a long way in building 'inclusive cities' (SDG 11), where women and other vulnerable groups enjoy equal access to social and economic opportunities in cities in SSA. By creating opportunities for women to participate in urban planning, their health and social needs and priorities with regard to housing and urban services will be prioritized. This will reduce socio-economic equalities and improve prospects for gender equality in urban areas (SDG 5). What is required is for both national and local governments to commit resources to implement some of these measures, as they will go a long way in addressing gender gaps in urban development and build inclusive and more prosperous cities that work for all.

References

Abrahams, N., Mathews, S., & Ramela, P. (2006). Intersections of 'sanitation, sexual coercion and girls' safety in schools. *Tropical Medicine & International Health, 11*(5), 751–756.

ActionAid. (2013). *Women and the city II: Combating violence against women and girls in urban in urban public spaces – The role of public services*. ActionAid.

Adama, O., & Lindell, I. (2020). *Visions of urban modernity and the shrinking of public space challenges for street work in African cities*. Uppsala.

Adams, E. A., & Nyantakyi-Frimpong, H. (2020). Stressed, anxious, and sick from the floods: A photovoice study of climate extremes, differentiated vulnerabilities, and health in old Fadama, Accra, Ghana. *Health & Place, 67*, 102500.

Agbadi, P., Darkwah, E., & Kenney, P. L. (2019). A multilevel analysis of regressors of access to improved drinking water and sanitation facilities in Ghana. *Journal of Environmental and Public Health, 2019*, 1.

Ajibade, I., McBean, G., & Bezner-Kerr, R. (2013). Urban flooding in Lagos, Nigeria: Patterns of vulnerability and resilience among women. *Global Environmental Change, 23*(6), 1714–1725.

Arku, G. (2009). Housing policy changes in Ghana in the 1990s. *Housing Studies, 24*(2), 261–272.

Arku, G., Luginaah, I., & Mkandawire, P. (2012). "You either pay more advance rent or you move out": Landlords/ladies' and tenants' dilemmas in the low-income housing market in Accra, Ghana. *Urban Studies, 49*(14), 3177–3193.

Awumbila, M. (2015). Women moving within borders: Gender and internal migration dynamics in Ghana. *Ghana Journal of Geography, 7*(2).

Awumbila, M., & Ardayfio-Schandorf, E. (2008). Gendered poverty, migration and livelihood strategies of female porters in Accra, Ghana. *Norwegian Journal of Geography, 62*(3), 171–179.

Azcona, G., Bhatt, A., Duerto, S., & Uteng, T. P. (2020). *Spotlight on goal 11. Harsh realities: Marginalized women in cities of the developing world*. UN Women.

Bass, L. E. (2000). Enlarging the street and negotiating the curb: Public space at the edge of an African *market. International Journal of Sociology and Social Policy, 20*(1/2), 74–95.

Beall, J. (1996). Participation in the city. *Gender and Development, 4*(1), 9–16. https://doi.org/10.1080/741921946

Bisung, E., & Elliott, S. J. (2016). 'Everyone is exhausted and frustrated': Exploring psychosocial impacts of the lack of access to safe water and adequate sanitation in Usoma, Kenya. *Journal of Water, Sanitation and Hygiene for Development, 6*(2), 205–214.

Bisung, E., & Elliott, S. J. (2017). Psychosocial impacts of the lack of access to water and sanitation in low-and middle-income countries: A scoping review. *Journal of Water and Health, 15*(1), 17–30.

Boamah, N. A. (2014). Housing policy in Ghana: The feasible paths. *Ghana Journal of Development Studies, 11*(1), 1–18.

Brouder, A., & Sweetman, C. (2015). Introduction: Working on gender issues in urban areas. *Gender and Development, 23*(1), 1–12.

Brown, R. (2010). Unequal burden: Water privatisation and women's human rights in Tanzania. *Gender and Development, 18*(1), 59–67.

Centre on Housing Rights and Evictions. (2008). *Women, slums and urbanisation: Examining the causes and consequences*. Women and Housing Rights Programme.

Chant, S. (2003). *Female household headship and the feminization of poverty: Facts, fictions and forward strategies*. New Working Paper Series. Issue 9. London: School of Economics. Gender Institute.

Chant, S. (2013). Cities through a 'gender lens': A golden 'urban age' for women in the global south? *Environment and Urbanization, 25*(1), 9–29.

Cities Alliance. (2020). *Planning for inclusive greater Banjul*. [Online] Available at: https://www.citiesalliance.org/sites/default/files/2021-01/Planning%20For%20Inclusive%20Greater%20Banjul_Cities%20Alliance.pdf. Accessed February 2021.

Corburn, J., & Hildebrand, C. (2015). Slum sanitation and the social determinants of women's health in Nairobi, Kenya. *Journal of Environmental and Public Health, 2015, 1*.

Escalante, S. O., & Valdivia, B. G. (2015). Planning from below: Using feminist participatory methods to increase women's participation in urban planning. *Gender and Development, 23*(1), 113–126. https://doi.org/10.1080/13552074.2015.1014206

Esquivel, V., & Sweetman, C. (2016). Gender and the sustainable development goals. *Gender and Development, 24*(1), 1–8.

Fainstein, S. S., & Servon, L. J. (2005). *Gender and planning: A reader*. Rutgers University Press.

Fenster, T. (2005). The right to the gendered city: Different formations of belonging in everyday life. *Journal of Gender Studies, 14*(3), 217–231. https://doi.org/10.1080/09589230500264109

Fisher, M. B., Shields, K. F., Chan, T. U., Christenson, E., Cronk, R. D., Leker, H., et al. (2015). Understanding handpump sustainability: Determinants of rural water source functionality in the greater Afram Plains region of Ghana. *Water Resources Research, 51*(10), 8431–8449.

Goebel, A. (2015). *On their own: Women, urbanization, and the right to the city in South Africa*. McGill-Queen's University Press.

Gomez, M., Perdiguero, J., & Sanz, A. (2019). Socio-economic factors affecting water access in rural areas of low- and middle-income countries. *Water, 11*(2), 202. https://doi.org/10.3390/w11020202

Horelli, L. (2017). Engendering urban planning in different contexts - successes, constraints and consequences. *European Planning Studies 25*(10), 1779–1796. https://doi.org/10.1080/09654313.2017.1339781

Institute for Women's Policy Research (IWPR). (2015). *Gender, urbanization and democratic governance*. National Democratic Institute.

International Finance Corporation. (2021). *Affordable housing in Africa*. [Online] Available at: Affordable Housing in Africa (ifc.org). Accessed Jan 2021.

International Organization for Migration. (2015). *World migration report: Migrants and cities, new partnerships to manage mobility*. IOM.

Kabeer, N. (2005). Gender equality and women's empowerment: A critical analysis of the third millennium development goal 1. *Gender and Development, 13*(1), 13–24. https://doi.org/10.1080/13552070512331332273

Karley, N. K. (2008). Ghana residential property delivery constraints and affordability analysis. *Housing Finance International, 22*(4), 22.

Kihato, C. W. (2011). The city from its margins: Rethinking urban governance through the everyday lives of migrant women in Johannesburg. *Social Dynamics, 37*(3), 349–362. https://doi.org/10.1080/02533952.2011.656432

Malaza, N. (2009). *Gender in planning and urban development*. Commonwealth Secretariat Discussion Paper. Organisation for Economic Co-operation and Development, Paris [Online]. Available at: http://www.oecd-ilibrary.org/content/ workingpaper/5k3w8fb9pbf1-en.

Marmot, M. (2014). *Status syndrome: How social standing affects our health and longevity*. Bloomsbury.

McFerson, H. (2010). Poverty among women in sub-Saharan Africa: A review of selected issues. *Journal of International Women's Studies, 11*(4), 50–72.

Meleis, A. I. (2011). Developing urban areas as if gender matters. In A. I. Meleis, E. L. Birch, & S. M. Wachter (Eds.), *Women's health and the world's cities*. University of Pennsylvania Press. https://doi.org/10.9783/9780812205084.

Meth, P. (2010). Unsettling insurgency: Reflections on women's insurgent practices in South Africa. *Planning Theory and Practice, 11*(2), 241–263. https://doi.org/10.1080/14649351003759714

Moser, C. (2012). Mainstreaming women's safety in cities into gender-based policy and programmes. *Gender and Development, 20*(3), 435–452. https://doi.org/10.1080/13552074.2012.731742

Moser, C., & McIlwaine, C. (2004). *Encounters with violence in Latin America*. Routledge.

Muller, M. (2008). Free basic water—A sustainable instrument for a sustainable future in South Africa. *Environment and Urbanization, 20*(1), 67–87.

Ngum, S. A. (2018). *Addressing gender issues in urban water and sanitation*. [Online] Available at: https://blogs.afdb.org/investing-in-gender-equality-for-africas-transformation/post/addressing-gender-issues-in-urban-water-and-sanitation-17741. Accessed Feb 2021.

Office of the High Commissioner for Human Rights. (2012). *Women and the right to adequate housing*. United Nations.

Oxfam. (2020). *Time to care: Unpaid and underpaid care work and the global inequality crisis*. Oxfam Briefing Paper.

Oxfam International. (2021). *Why the majority of the world's poor are women?* [Online] Available at: Why the majority of the world's poor are women Oxfam International. Accessed Jan 2021.

Peprah, D., Baker, K. K., Moe, C., Robb, K., Wellington, N. I. I., Yakubu, H., & Null, C. (2015). Public toilets and their customers in low-income Accra, Ghana. *Environment and Urbanization, 27*(2), 589–604.

Razavi, S., & Turquet, L. (2016). Progress of the world's women 2015–2016: Transforming economies, realizing rights. *Global Social Policy, 16*(1), 86–93.

Saghir, J., & Santoro, J. (2018). *Urbanization in sub-Saharan Africa*. Meeting Challenges by Bridging Stakeholders. Centre for Strategic and International Studies [Online] www.csis.org. Accessed Feb 2021.

Sánchez de Madariaga, I., & Neuman, M. (2016). Mainstreaming gender in the city. *Town Planning Review, 87*(5), 493–504. https://doi.org/10.3828/tpr.2016.33

Sorenson, S. B., Morssink, C., & Campos, P. A. (2011). Safe access to safe water in low-income countries: Water fetching in current times. *Social Science & Medicine, 72*(9), 1522–1526. https://doi.org/10.1016/j.socscimed.2011.03.010

Tufuor, T. (2009). *Gender and women housing problems in Accra–the case of old Fadama*. Ghana Ministry of Water Resources Works and Housing.

Tusting, L. S., Bisanzio, D., Alabaster, G., Cameron, E., Cibulskis, R., Davies, M., et al. (2019). Mapping changes in housing in sub-Saharan Africa from 2000 to 2015. *Nature, 568*(7752), 391–394.

UN Women. (2018). *Turning promises into action: Gender equality in the 2030 agenda*. UN Women.

UN-Habitat. (2011). *Affordable land and housing in Africa*. United Nations Human Settlements Programme.

UN-Habitat. (2015a). *Habitat III policy paper framework, 1 right to the city and cities for all*. [Online] http://www.europeanhabitat.com/wp-content/uploads/2016/03/PU1-Right-to-the-City.pdf.

UN-Habitat. (2016). *New urban agenda*. IN The United Nations Conference on Housing and Sustainable Urban Development Habitat III.

UN-Habitat, UNCDF, UCLG-Africa, UNECA. (2020). *COVID-19 in African cities: Impacts, responses and policies*. [Online] Available at: final_-_covid_in_african_cities_9_june_2020_.pdf (unhabitat.org). Accessed Jan 2020.

UNICEF. (2017). *Gender-responsive water, sanitation and hygiene: Key elements for effective wash programming*. Retrieved from https://www.unicef.org/gender/files/Gender_Responsive_WASH.pdf. Accessed 04 Jan 2021.

United Nations. (2015a). *Transforming our world: The 2030 agenda for sustainable development, United Nations, A/RES/70/1*. [Online] Available at: from https://sustainabledevelopment.un.org/

content/documents/21252030%20Agenda%20for%20Sustainable%20Development%20web.pdf. Accesssed Jan 2021.

United Nations. (2015b). *Transforming our world: The 2030 agenda for sustainable development.* Retrieved from https://sustainabledevelopment.un.org/content/documents/21252030%20Agenda%20for%20Sustainable%20Development%20web.pdf. Accessed 04 Jan 2021.

United Nations. (2017). *The sustainable development goals report 2017.* United Nations.

United Nations Department of Economic and Social Affairs, Population Division. (2019). *World urbanization prospects: The 2018 revision. (ST/ESA/SER.A/420).* United Nations.

United Nations Development Programme (UNDP). (2006). *Beyond scarcity: Power, poverty, and the global water crisis: UN Human Development Report 2006.* New York: United Nations Development Programme, Human Development Report Office.

UN-Women. (2013). *Tanzanian Women bring safe drinking water to their communities.* Retrieved from https://www.unwomen.org/en/news/stories/2013/3/tanzanian-water-kiosks. Accessed 17 Jan 2021.

UN-Women. (2020). Harsh realities: Marginalized women in cities of the developing world. *Spotlight on Goal 11.* Retrieved from https://sustainabledevelopment.un.org/content/documents/2700harshrealitiesmarginalizedwomenincitiesofthedevelopingworlden.pdf.

Venter, A., & Marais, L. (2006). Gender and housing policy in South Africa: Policy and practice in Bloemfontein. *Journal of Family Ecology and Consumer Sciences, 34*, 69–79.

Venter, C., Vokolkova, V., & Michalek, J. (2007). Gender, residential location, and household travel: Empirical findings from low-income urban settlements in Durban, South Africa. *Transport Reviews, 27*(6), 653–677.

Water and Sanitation Program. (2010). *Mainstreaming gender in water and sanitation: Gender in water and sanitation.* World Bank.

Whitzman, C., Legacy, C., Andrew, C., Klodawsky, F., Shaw, M., & Viswanath, K. (Eds.). (2013). *Building inclusive cities: Women's safety and the right to the city.* Routledge.

WHO/UNICEF. (2013). *Progress on sanitation and drinking water: 2013 update report.* World Health Organisation and the United Nations Children's Fund (UNICEF).

World Bank. (2006). Gender, Time Use, and Poverty in Sub- Saharan Africa World Bank Working paper no. 73, 2006, The International Bank for Reconstruction and Development, Washington, D.C: World Bank.

World Bank. (2015a). *Community-led partnerships for resilience.* [Online] Available at: from https://www.gfdrr.org/sites/default/files/publication/Community_led_partnership_JUNE24.pdf. Accessed Feb 2021.

World Bank. (2015b). *Stocktaking of the housing section in sub-Saharan Africa.* World Bank.

World Bank. (2019). *Urban population growth rate.* [Online] Urban population growth (annual %)-Sub-Saharan Africa | Data (worldbank.org). Accessed Feb 2021. World Bank Group.

World Bank. (2020). *Sub-Saharan Africa* [Online] Available at: Global_POVEQ_SSA.pdf (worldbank.org). Accessed Feb 2020.

World Health Organisation. (2011). Guidelines for drinking-water quality. *WHO Chronicle, 38*(4), 104–108.

World Health Organization and the United Nations Children's Fund (2017). Progress on Drinking Water, Sanitation and Hygiene: 2017 Update and SDG Baseline. Geneva: Switzerland

Yon, A., & Nadimpalli, S. (2017). Cities for whom? Re-examining identity, to reclaim the right to the city for women. *Australian Planner, 54*(1), 33–40. https://doi.org/10.1080/07293682.2017.1297317

Zack, T., Charlton, S., & De Villiers, R. (2003). *Better off, but...: Beneficiaries' perceptions of the Government's housing subsidy scheme.* Housing Finance Resource Programme.

Zimbabwe Women Resources Centre and Network. (2015). *Gender responsive budgeting in selected local authorities in Zimbabwe.* [Online] https://www.zwrcn.org.zw/index.php/our-work/geppf. Accessed Feb 2020.

Part IV
Infectious Diseases and Pandemic

Chapter 11
Participation in Social Activities After Ebola Virus Disease Outbreak in Sierra Leone: Does Where You Live Matter?

Joseph Kangmennaang, Medjatu Kuyateh, and Nasong A. Luginaah

Introduction

Emerging and re-emerging infectious disease outbreaks are associated with considerable fear and panic among the general public. Fear and stigma associated with diseases such as severe acute respiratory syndrome (SARS), Influenza, Ebola virus disease (EVD), and the COVID-19 pandemic illustrate some of these difficulties survivors must endure. Such fear and stigma are due to the evolving nature and complex uncertainties associated with these outbreaks. Mitigating fear and discrimination directed toward persons infected with and/or affected by infectious diseases is important in controlling transmission and promoting the well-being of infected and affected persons, their households, and local communities. In some instances, persons who are stigmatized may delay seeking care and remain in the community undetected.

EVD was first discovered in Zaire (*now the Democratic Republic of Congo (DRC)*) in 1976. Ever since, there have been periodic outbreaks within the DRC and surrounding regions (WHO, 2020). The 2014–2016 EVD outbreak in West Africa was considered an unprecedented public health emergency in the history of the

J. Kangmennaang (✉)
School of Kinesiology & Health Studies, Queen's University, Kingston, ON, Canada
e-mail: j.kangmennaang@queensu.ca

M. Kuyateh
Department of Public Health Sciences, College of Health and Human Services, University of North Carolina at Charlotte, Charlotte, NC, USA

N. A. Luginaah
Schulich School of Medicine and Dentistry, Faculty of Public Health, Western University, London, ON, Canada
e-mail: nluginaa@uwo.ca

J. A. Braimah et al. (eds.), *Health Geography in Sub-Saharan Africa*, Global Perspectives on Health Geography, https://doi.org/10.1007/978-3-031-37565-1_11

disease (WHO, 2016a). About 11,322 persons died due to EVD, and estimates put the number of EVD survivors at more than 10,000 (WHO, 2016). Most EVD survivors are known to suffer from short and long-term physical symptoms and mental complications (Lotsch et al., 2017; Vetter et al., 2016; James et al., 2020). Psychosocial consequences of EVD survivorship can be distressful, in part due to the adverse experiences of individuals and EVD impacted households during infection, treatment, or after treatment. These adverse experiences may include various forms of psychosocial health challenges such as depression, fear, anxiety, grief due to the loss of loved ones, and stigma (Lotsch et al., 2017, James et al., 2020).

EVD-related stigma has been reported by EVD survivors and their communities in DR Congo (35%), Guinea (26%), and Liberia (3%) (De Roo et al., 1998; de St. Maurice et al., 2018; Etard et al., 2017), and is more common among female than male EVD survivors (Hewlett & Amola, 2003). Anecdotal evidence indicates that some individuals cured of EVD were shunned by their communities and prohibited from returning to their homes. This abandonment only prolongs the effects of fear and stigma as affected people, and places may become permanently labelled. Hence, there have been calls for more research on how to reduce stigma from highly fatal infectious diseases because stigmatization tends to discourage those with the disease from reporting and seeking care (Obilade, 2015; James et al., 2020). The 2020 COVID-19 epidemic has only reinforced the need to understand the impacts of stigma during epidemics. There is no doubt that the fear and rumors that tend to emerge from epidemics with high mortality rates, such as Ebola or COVID-19, can become a vicious cycle. Among the published studies on EVD-related stigma, none have explored the socio-demographic and place-based determinants among affected communities in the general population (James et al., 2020). Therefore, the aim of this study is to estimate the social determinants of non-participation in social life after the EVD epidemic outbreak in Sierra Leone. Understanding the outbreak's non-EVD health impacts and pathways is important for rebuilding the health system in affected countries and improving response to future health crises.

Study Context

Sierra Leone is a sub-Saharan African country located in West Africa and bordered by Liberia to the southeast, Guinea to the northeast, and the Atlantic Ocean to the south. Sierra Leone has a total land area of 71,740 km^2 with a population of 8.4 million (World Bank, 2021). The country is divided into four regions, Northern Province, Southern Province, the Eastern Province, and the Western Area, which represent the second level of administrative division (Fig. 11.1). After going through a terrible civil war from 1991 to 2002, the Sierra Leonean health system was beginning to recover from the effects of the war (Christensen et al., 2021) when the 2014 Ebola epidemic overwhelmed the healthcare system, with anecdotal reports indicating that the epidemic may have eroded what was already fragile. The magnitude and rate of transmission of the 2014 EVD outbreak presented a significant

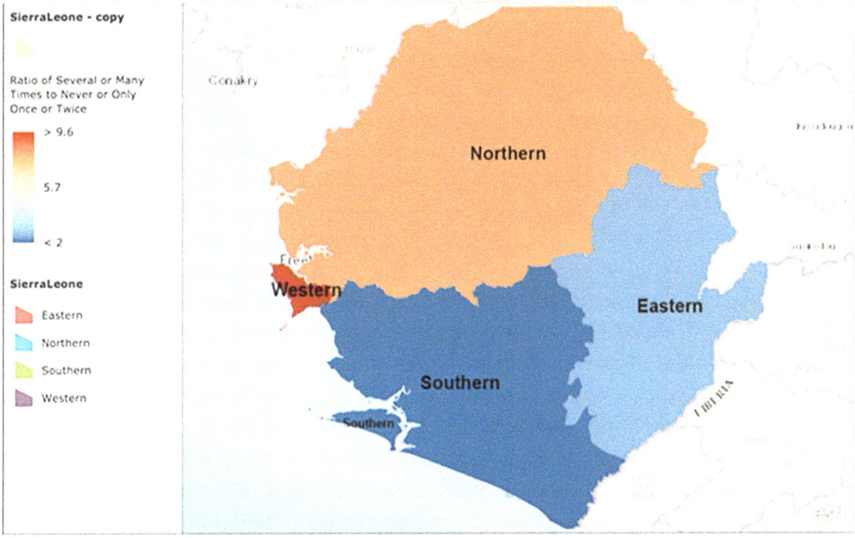

Fig. 11.1 Map of Sierra Leone showing relative levels of Ebola stigma

threat to the Sierra Leonean public health system and health infrastructure due to the scarcity of healthcare workers and lack of basic healthcare resources such as personal protective equipment and guidelines for frontline workers in the early phase of the epidemic (Boozary et al., 2014; Christensen et al., 2021). The country has 2.2 healthcare workers per 10,000 and approximately 4 hospital beds per 10,000 residents. In response to the outbreak, the United Nations Security Council unanimously passed a resolution to establish the United Nations Mission for Emergency Ebola Response (UNMEER), the first of its kind (Shoman et al., 2017).

Methods

Data for this cross-sectional analysis was drawn from the sixth wave of the Sierra Leone Afrobarometer (SLA) survey. The Afrobarometer survey is a nationally representative household survey of public attitudes toward governance, economic, and other political and social issues. The surveys have been conducted in more than 30 African countries since 1999 (Afro Barometer, 2020). The SLA survey captures exposure to EVD as well as individual, household, and community characteristics using a clustered, stratified, multi-stage proportion probability sampling procedure that is designed to generate samples that are nationally representative of voting age citizens. Sampling units were first randomly selected from the 2015 Sierra Leonean Population and Housing Census and then within sampling units, households were randomly selected for participation in the SLA survey. In sampled households, enumerators alternate between interviewing a voting age male or female adult within each household in order to generate gender balance among survey respondents. In the sixth round, 1191 respondents were interviewed in Sierra Leone. Surveys were planned for all three of the heavily affected countries before the Ebola outbreak but were not fielded until the outbreaks were largely controlled in 2015. Additional questions were added to the survey to ask respondents about their experiences during the Ebola outbreak. Specifically, respondents were asked, "*how often, if ever, have you or anyone in your family unable to; (1) attend school; (2) go to work or other income-generating activities; (3) attend social gatherings and other communal events, and (4) unable to get medical care for other health problems due to Ebola?* Other questions included: *"Do you know a close friend or relative who was infected with Ebola?"* with *yes, no, refused,* and *don't know* as potential answers. Regardless of their answer to the first question, the survey asked, *"Do you know a close friend or relative who died of Ebola?"* with the same set of potential answers. Field dates for the survey were May 22–June 10, 2015. By May 27, 2015, in Sierra Leone almost 90% of eventual cases were already reported (12,706 of 14,124), representing the time periods toward the tail end of the outbreaks (Grépin et al., 2020). Individual-level characteristics drawn from the survey included the age of the respondent, gender, employment status, religion, and the level of education completed (no formal education, some primary education, some secondary education, and some post-secondary education). Household-level variables were constructed in terms of whether the household was in an urban area as well as region of residence. We also used data from the Afrobarometer to construct household and neighborhood-level wealth indices at the sampling unit level as described in the following text. Data on the administrative level unit or region of the household and neighborhood level variables were also obtained from the Afrobarometer survey. There are 4 first-level administrative units in Sierra Leone. We refer to these as regions in the rest of the paper.

Dependent variable: We defined severity of participation in social life as a binary variable, which we defined as '1' for all those who reported not participating in social life several or many times and '0' for those who were able to participate in social life once or twice.

Independent variables: We defined an EVD exposure as a binary variable constructed from two related questions, which were defined as 1 if the survey respondent reported knowing one or more close friends or family members who had Ebola or 0 otherwise, and the second which was coded as 1 if the respondent knew a friend who died of Ebola or 0 otherwise. Like Grépin et al. (2020), we used principal component analysis (PCA) to construct two measures of wealth-related socio-economic status (SES) at the household- and neighborhood-levels. At the household-level, the wealth index was built based on ownership of the following assets or goods: shelter type, roofing material, water source, toilet location, radio, television, vehicle, mobile phone, remittance receipt, access to food and water, access to medications, fuel, and cash. We first calculated the wealth index using all these items and then calculated the wealth index quintiles. A separate neighborhood-level SES index was constructed using PCA at the sampling unit-level resources. The neighborhood level SES was constructed using responses to questions about access to these items at the sampling unit level: passable roads, paved roads, electricity grid, piped water, mobile phone service, post office, school, police station, health clinic, market stalls, bank, and public transportation. We created a binary neighborhood inclusivity variable as an additive variable from responses to five questions that asked about acceptability of people from different religions, ethnicities, homosexuals, people living with HIV/AIDS, immigrants, and foreign workers. We also adjusted models for demographics (urban residence, age of the respondent, the number of adults in the respondent's household, and gender of the respondent).

Results

Table 11.1 shows the descriptive statistics for the sample. Notably, 36% of respondents had been exposed to EVD; either they knew someone who was infected or died of EVD and 71% of this population reported not participating in social life. Also, 61% of respondents believed that the Sierra Leonean Government effectively responded to the EVD outbreak. The mean age of the sample was 40 years [range 18–99 years] and about half of the respondents were male. The average household size was 7 persons [range 1–35]. Most respondents reported secondary education (35%) as their highest level of education while about 40% of respondents indicated they were currently employed. Most of the respondents (41%) lived in neighborhoods with medium SES, and approximately 77% of respondents lived in inclusive neighborhoods.

In Table 11.2, we present the determinant of non-participation results along individual, household, and neighborhood lines. The dependent variable in all our models is a binary variable reporting non-participation and the models are all estimated using binary logistic regression. The first set of columns corresponds to the results controlling for only individual and demographic factors, while the second columns control for individual, household, and community level factors.

In model 1, after controlling for individual and demographic factors, we find that respondents who reported EVD exposure (OR = 1.39, $p \leq 0.05$) were more likely to report non-participation than those who were not exposed to EVD. Employed individuals (OR = 1.88, $p \leq 0.01$) were also more likely to report non-participation compared to the unemployed. In model 2, after controlling for household and community level factors, the association between EVD exposure and non-participation became stronger. We also found that compared to the East, respondents in the Northern (OR = 1.62, $p \leq 0.01$) and Western (OR = 1.87, $p \leq 0.01$) regions were more likely to report non-participation, while respondents in the South (OR = 0.74, $p \leq 0.05$) were less likely to report non-participation. Similarly, compared to urban areas, residents in rural areas (OR = 0.75, $p \leq 0.05$) were less likely to report non-participation. Neighborhood SES also predicted non-participation, with medium (OR = 0.68, $p \leq 0.01$) and high (OR = 0.52, $p \leq 0.01$)

Table 11.1 Descriptive statistics of respondents in Sierra Leone

Variable	Category	Frequency(%)
Non-participation in social life	Never	253 (21.24)
	Once or twice	87(7.38)
	Several times	279 (23.43)
	Many times	572 (48.03)
Unable to work	Not often Often	330 (27.71) 861(72.29)
Unable to access non-EVD medical care	Not often Often	336 (28.21) 855(71.79)
Unable to attend social gatherings	Not often Often	290 (24.35) 901(75.65)
Unable to attend school	Not often Often	635 (52.96) 564(47.04)
Ebola affected		
No		765(64.23)
Yes		426(35.77)
Effectiveness of Ebola response		
	Not effective	460(36.62)
	Effective	731(61.38)
Community inclusiveness		
	No	266(22.33)
	Yes	925(77.67)
Neighborhood socio-economic status		
	Low	436(36.61)
	Middle	493(41.39)
	High	262(22.00)
Employment status		
	Unemployed	719(60.37)
	Employed	472(39.63)
Educational level		
	Never	547(45.93)
	Primary Secondary Tertiary	159(13.35) 419(35.18) 66(5.54)
Gender		
	Male	589(49.45)
	Female	602 (50.55)
Religion		
	Christianity Islam	273(23.17) 885(74.31)
	Others	30(2.52)
Area of residence		
	Urban	443 (37.20)
	Rural	748 (62.80)

(continued)

Table 11.1 (continued)

Variable	Category	Frequency(%)
Region of residence		
	East	301(25.27)
	North	392(32.91)
	South	260(21.87)
	Western	238(19.98)
Asset levels		
	Poor	425(35.68)
	Middle	312(26.20)
	Rich	454(38.12)
Age of respondent		
	18–30	521(43.74)
	31–60	555(46.60)
	>60	115(9.66)
Observations		**1191**

SES neighborhoods being less likely to experience non-participation compared to low SES neighborhoods. Respondents who believe that the government EVD response was effective (OR = 0.66, $p \leq 0.01$) were less likely to report high levels of non-participation. On the contrary, respondents in neighborhoods that are inclusive (OR = 1.44, $p \leq 0.01$) were more likely to report EVD stigma compared to non-inclusive neighborhoods.

Determinants of the Different Components of Participating in Social Activities

Non-Ebola Health Care Access

We present the determinants of the different non-participation components in Table 11.3. The determinants of non-Ebola medical related care include (see model 1) EVD exposure, region of residence, neighborhood SES, neighborhood inclusiveness, and effectiveness of EVD response. Respondents who reported EVD exposure (OR = 2.22, $p \leq 0.01$) were more likely to report inability to access non-Ebola health care than those not exposed to EVD. Compared to the East, respondents in the Northern (OR = 6.77, $p \leq 0.01$) and Western (OR = 10.79, $p \leq 0.01$) regions were more likely to report inability to access non-Ebola medical care. Also, respondents living in neighborhoods with medium (OR = 0.43, $p \leq 0.05$) and high (OR = 0.31, $p \leq 0.05$) levels of SES were less likely to report inability to access non-Ebola medical care than respondents in low SES neighborhoods. However, respondents living in inclusive neighborhoods (OR = 1.64, $p \leq 0.01$) were more likely to report inability to non-Ebola medical care than respondents in non-inclusive neighborhoods.

Table 11.2 Determinants of non-participation in social life after Ebola outbreak in Sierra Leone

Independent variables	OR(95%)	OR(95%)
Non-participation in social life (ref: No)	Model 1	Model 2
Yes	1.39(1.02–1.89)**	1.31(1.11–1.55)***
Gender of respondent (ref: Male)		
Female	0.97(0.77–1.23)	0.95(0.812–1.12)
Age category (ref: 18–35)		
36–60	0.90(0.70–1.16)	0.90(0.76–1.07)
60+	0.68(0.43–1.07)*	0.68(0.51–0.91)**
Household size	1.02(0.97–1.05)	1.01(0.98–1.03)
Religion (ref: Christianity)		
Muslim	1.12(0.84–1.50)	1.07(0.885–1.31)
Others	2.28(0.85–6.15)	1.26(0.71–2.22)
Educational level (ref: No education)		
Primary	1.40(0.92–2.11)	1.26(0.98–1.63)*
Secondary	1.05(0.78–1.40)	1.02(0.84–1.24)
Tertiary	0.55(0.30–1.02)*	0.76(0.51–1.11)
Employment status (ref: Unemployed)		
Employed	1.88(1.20–2.94)***	1.28(1.08–1.52)***
Asset level (ref: Low)		
Medium	0.82(0.60–1.12)	0.79(0.64–0.98)**
High	1.37(0.96–1.95)*	0.98(0.77–1.24)
Community level factors		
Region of residence (ref: East)		
North		1.62(1.28–2.04)***
South		0.74(0.57–0.94)**
Western		1.87(1.42–2.46)***
Area of residence (ref: Urban)		
Rural		0.75(0.59–0.96)**
Neighborhood socio-economic status (ref: Low)		
Middle		0.68(0.55–0.82)***
High		0.52(0.39–0.69)***
Effectiveness of Ebola response (ref: Not effective)		
Effective		0.88(0.74–1.04)
Neighborhood inclusiveness (ref: No)		
Yes		1.44(1.18–1.74)***
Constant		1.50(0.93–2.42)*
Observations	**1191**	**1191**

Notes: *OR* odds ratio, *Ref* Reference categories, *CI* confidence intervals
*$p \leq 0.10$, **$p \leq 0.05$, ***$p \leq 0.01$

Table 11.3 Determinants of Ebola-related stigma in Sierra Leone

Independent variables	Unable to access non-EVD medical care	Unable to attend social gatherings	Unable to work	Unable to attend school
Non-participation in social life (ref: No)	OR(95%)	OR(95%)	OR(95%)	OR(95%)
Yes	2.22(1.16–4.21)**	1.45(1.02–2.05)**	1.79(1.23–2.61)***	1.17(0.85–1.60)
Gender of respondent (ref: Male)				
Female	1.10(0.71–1.69)	0.90(0.67–1.23)	0.95(0.71–1.27)	1.12(0.83–1.51)
Age category (ref: 18–35)				
36–60	0.75(0.47–1.20)	0.95(0.69–1.30)	0.85(0.62–1.17)	0.91(0.67–1.23)
60+	0.66(0.30–1.43)	0.56(0.32–0.99)**	0.52(0.29–0.91)**	0.63(0.37–1.08)*
Household size	1.04(0.96–1.12)	1.03(0.98–1.08)	1.03(0.98–1.07)	1.00(0.95–1.04)
Religion (ref: Christianity)				
Muslim	0.87(0.50–1.50)	1.08(0.74–1.57)	1.04(0.72–1.49)	1.15(0.80–1.65)
Others	1.30(0.28–5.93)	1.84(0.54–6.16)	1.88(0.60–5.85)	0.68(0.26–1.79)
Educational level (ref: No education)				
Primary	1.28(0.65–2.50)	1.51(0.91–2.49)	1.51(0.92–2.45)*	1.38(0.87–2.20)
Secondary	1.30(0.77–2.18)	1.13(0.77–1.65)	1.00(0.69–1.43)	1.26(0.87–1.82)
Tertiary	0.49(0.17–1.43)	0.63(0.30–1.33)	0.47(0.22–0.99)**	0.50(0.24–1.03)*
Employment status (ref: Unemployed)				
Employed	2.35(0.97–5.65)*	1.70(1.15–2.51)***	1.83(1.24–2.69)***	1.56(1.09–2.22)**
Asset level (ref: Low)				
Medium	0.60(0.30–1.22)	0.67(0.44–1.02)*	0.81(0.55–1.20)	0.62(0.42–0.94)**
High	0.70(0.36–1.35)	0.92(0.59–1.44)	0.85(0.54–1.32)	0.80(0.52–1.24)
Region of residence (ref: East)				
North	6.77(2.62–17.49)***	1.75(1.06–2.87)**	2.09(1.24–3.52)***	2.49(1.45–4.28)***
South	0.56(0.13–2.29)	0.40(0.23–0.69)***	0.48(0.29–0.81)***	0.65(0.41–1.02)*
Western	10.79(2.87–40.55)***	3.27(1.56–6.84)***	2.94(1.50–5.75)***	4.30(1.94–9.53)***

(continued)

Table 11.3 (continued)

Independent variables	Unable to access non-EVD medical care	Unable to attend social gatherings	Unable to work	Unable to attend school
Area of residence (ref: Urban)				
Rural	0.76(0.39–1.48)	0.73(0.46–1.16)	0.83(0.53–1.29)	0.53(0.32–0.86)**
Neighborhood socio-economic status (ref: Low)				
Middle	0.43(0.21–0.87)**	0.60(0.40–0.91)**	0.58(0.39–0.88)**	0.61(0.41–0.90)**
High	0.31(0.12–0.79)**	0.39(0.21–0.74)***	0.52(0.29–0.90)**	0.42(0.23–0.77)***
Effectiveness of Ebola response (ref: Not effective)				
Effective	0.67(0.40–1.13)	0.87(0.62–1.21)	0.79(0.57–1.11)	0.92(0.67–1.25)
Neighborhood inclusiveness (ref: No)				
Yes	4.02(1.87–8.61)***	2.76(1.64–4.66)***	2.67(1.60–4.45)***	2.83(1.68–4.76)***
Constant	0.71(0.19–2.55)	1.18(0.47–2.94)	0.85(0.35–2.06)	1.04(0.43–2.47)
Observations	**1191**	**1191**	**1191**	**1191**

Notes: *OR* odds ratio, *Ref* Reference categories, *CI* confidence intervals
*$p \leq 0.10$, **$p \leq 0.05$, ***$p \leq 0.01$

Participation in Social Gathering

In model 2 of Table 11.3, the results indicate that Ebola exposure, older age, employment status, region of residence, neighborhood SES, and neighborhood inclusiveness were significant predictors of participants' inability to attend social gathering due to EVD. Respondents who reported EVD exposure (OR = 1.45, $p \leq 0.05$) were more likely to report inability to participate in social gatherings due to EVD than their counterparts who were not exposed to EVD. Also, respondents living in medium (OR = 0.60, $p \leq 0.05$) and high (OR = 0.39, $p \leq 0.01$) SES neighborhoods were less likely to report inability to participate in social gatherings due to EVD compared to poor SES neighborhoods. Compared to the East, respondents in the Northern (OR = 1.75, $p \leq 0.05$) and Western (OR = 3.27, $p \leq 0.01$) regions were more likely to report participation in social gatherings while southern residents (OR = 0.40, $p \leq 0.01$) were less likely to express these concerns.

Work-Related Participation

In model 3 of Table 11.3, the results indicate that exposure to EVD, age, educational level, employment status, region of residence, and neighborhood SES significantly predicted respondents' inability to work or access work due to concerns with EVD. Respondents who reported EVD exposure (OR = 1.79, $p \leq 0.01$) were more likely to report inability to work due to EVD than those not exposed to EVD. Also, participants in medium (OR = 0.58, $p \leq 0.05$) and high (OR = 0.52, $p \leq 0.05$) SES neighborhoods were less likely to report inability to work due to EVD compared to respondents in low SES neighborhoods.

School Attendance

In model 4 of Table 11.3, the results indicate that residing in the Northern (OR = 2.49, $p \leq 0.01$) and Western (OR = 4.30, $p \leq 0.01$) regions, rural residence (OR = 0.53, $p \leq 0.05$), currently employed (OR = 1.56, $p \leq 0.05$), medium (OR = 0.62, $p \leq 0.01$), and high (OR = 0.46, $p \leq 0.01$) neighborhood SES were significantly associated with respondents' inability to attend school due to EVD.

Discussion

The purpose of this study was to investigate determinants of participation in social life in Sierra Leone, following the 2014 Ebola epidemic in West Africa. Consistently, more than 50% of respondents indicated their inability to work, inability to access non-EVD medical care, and inability to attend social gatherings. These findings are consistent with those reported among Ebola survivors from longitudinal and cross-sectional studies in the months after discharge from Ebola treatment unit (Deen et al., 2015; Overholt et al., 2018; Rabelo et al., 2016). For instance, many studies found elevated levels of distress, fear, anxiety, and depression among Ebola survivors. The pervasiveness and intensity of stigma would have both emotional and physical consequences. Following the Ebola epidemic in Sierra Leone, some individuals cured of EVD with certification of clearance by medical authorities were still prohibited from returning to their homes, and shunned by their communities, friends, family, and co-workers (Davtyan et al. 2014; Lasuta, 2014). In addition to these hostilities, other survivors have also reported threats of violence from their community (Davtyan et al. 2014; Lasuta, 2014). These social pressures and neglect help to prolong the effects of potential fear and stigma as these individuals and their area of residence can be permanently labelled. Hence, understanding the prevalence and social determinants of non-participation in social life is important to retool and redesign health systems to respond to emotional stresses caused by epidemics and pandemics.

Overall, almost 60% of households across all regions reported elevated levels of non-participation (composite measure), yet with inherent geographical variations. We found interesting regional differences as approximately 92%, 89%, 72%, and 59% of residents in the Western, Northern, Eastern, and Southern regions, respectively, reported non-participation. This is not surprising as the Western, Northern, and Eastern parts of the country were severely impacted by the 2014 Ebola outbreak. This regional variation draws attention to a need to understand how broader contextual factors such as culture, values, rurality, and other socio-economic factors may be influencing stigma, fear, and the inability of survivors to return to normal life.

The regional variations in non-participation showed that the Western region had the highest levels of non-participation, and this points to the importance of identifying and acknowledging places that are less likely to "return to normal" after a pandemic or epidemic, and their relational aspects with other places. Consistently, we found a non-participation-SES gradient where higher SES communities were less likely to report non-participation compared to low SES neighborhoods. A combination of poverty and non-participation may result in neighborhoods being seen as "unhealthy places" that are already suffering from the lack of social amenities such as inadequate infrastructure, roads, and health care services. These kinds of unhealthy associations need to be unpacked using place-based analysis to inform public health interventions (Smith-Morris, 2017). The resulting discrimination that may result from labelling such places may either result in underestimating or overestimating stigma (Overholt et al., 2018; Reid et al., 2014; Vorauer, 2006).

In Sierra Leone, the Ebola epidemic has left 'scared communities' characterized by stressful chronic interactions between individuals and groups within a community as well as with outsiders. Given the unequal impact of EVD on different neighborhoods (Luginaah, et al., 2016), this may instigate or reinforce group differences thus propagating inter-community hostilities. These findings highlight the need to adopt multi-level and community-based strategies in responding to epidemics such as Ebola Zika, SARs, and the more recent COVID-19 pandemic (Abramowitz et al., 2015; Konkor et al. 2020; Tenkorang, 2017). The effects of epidemics are often worsened and reinforced by broader structural inequalities and systemic discrimination (Konkor et al. 2020). Such structural factors individually and collectively shape human vulnerability to disease and its distribution and dictate attitudes, beliefs, and behaviors toward certain groups (Davtyan et al. 2014). As the transmission of emerging and re-emerging infectious diseases such as EVD, Zika, and COVID-19 intensify across the world, interventions aimed at raising community awareness and changing attitudes and values by providing correct and evidenced-based information may help reduce the labelling of persons and communities.

At the individual level, factors such as exposure to EVD, age, employment level, and household wealth were associated with the composite non-participation and the individual components of non-participation. Other important individual-level factors that predicted non-participation included employment status and household wealth.

Limitations

While our study may have helped to shed some light on participation in social activities after an epidemic outbreak, it has limitations that must be taken into consideration when interpreting the results. First, our measure of participation might reflect the unintended effects of restrictions and pandemic control measures that were implemented in response to the epidemic. Second, due to the cross-sectional nature of our data, we limited our interpretations to associations and refrained from making causal inferences. Third, because the data was collected almost a year into the Ebola epidemic, individual exposure and stigma perceptions may be overstated or understated. As with all survey-based research, responses to questions may also be subject to social desirability and recall bias. Notwithstanding these limitations, this study is one of the first to provide compelling evidence of EVD-related stigma among the general population, with emphasis on community-level factors and stigma-by-association.

Conclusion

The various infectious diseases outbreaks, notably the West African outbreak of EVD, and the current COVID-19 pandemic is a wake-up call to the global health community to do a better job at not only preventing and mitigating the spread of infectious diseases, but also understanding long-term impacts of these epidemics and pandemics on affected persons and communities. Current evidence indicates that fear and stigma can undermine preventive efforts by excluding individuals, families, and communities. This paper has broadened the literature on the psychosocial aspects of epidemics by looking at the social determinants of participating in social life after an epidemic outbreak. Our findings demonstrate convincing evidence that both individual and community-level interventions are needed to reduce Ebola-related stigma. These findings highlight the need to adopt multi-level and community-based strategies in responding to the aftermaths of disease outbreaks such as Ebola.

References

Abramowitz, S. A., McLean, K. E., McKune, S. L., Bardosh, K. L., Fallah, M., Monger, J., et al. (2015). Community-centered responses to Ebola in urban Liberia: The view from below. *PLoS Neglected Tropical Diseases, 9*(4).

Afro Barometer. 2020: *Afrobarometer data*. Retrieved from https://www.afrobarometer.org/data/. Accessed 04 Dec 2021.

Boozary, A. S., Farmer, P. E., & Jha, A. K. (2014). The Ebola outbreak, fragile health systems, and quality as a cure. *JAMA, 312*(18), 1859–1860.

Christensen, D., Dube, O., Haushofer, J., Siddiqi, B., & Voors, M. (2021). Building resilient health systems: Experimental evidence from Sierra Leone and the 2014 Ebola outbreak. *The Quarterly Journal of Economics, 136*(2), 1145–1198.

Davtyan, M., Brown, B., & Folayan, M. O. (2014). Addressing Ebola-related stigma: Lessons learned from HIV/AIDS. *Global Health Action, 7*(1), 26058.

De Roo, A., Ado, B., Rose, B., Guimard, Y., Fonck, K., & Colebunders, R. (1998). Survey among survivors of the 1995 Ebola epidemic in Kikwit, Democratic Republic of Congo: Their feelings and experiences. *Tropical Medicine & International Health, 3*(11), 883–885.

de St. Maurice, A., Ervin, E., Orone, R., Choi, M., Dokubo, E. K., Rollin, P. E., … & Fankhauser, J. (2018, October). Care of Ebola survivors and factors associated with clinical sequelae—Monrovia, Liberia. In *Open forum infectious diseases* (vol. 5, no. 10, pp. ofy239). Oxford University Press.

Deen, G. F., Knust, B., Broutet, N., Sesay, F. R., Formenty, P., Ross, C., Thorson, A. E., Massaquoi, T. A., Marrinan, J. E., & Ervin, E. (2015). Ebola RNA persistence in semen of Ebola virus disease survivors preliminary report. *New England Journal of Medicine, 377*, 1428–1437.

Etard, J. F., Sow, M. S., Leroy, S., Toure, A., Taverne, B., Keita, A. K., Msellati, P., Magassouba, N., Baize, S., Raoul, H., et al. (2017). Multidisciplinary assessment of post- Ebola sequelae in Guinea (Postebogui): An observational cohort study. *The Lancet Infectious Diseases, 17*(5), 545–552.

Grépin, K. A., Poirier, M. J., & Fox, A. M. (2020). The socio-economic distribution of exposure to Ebola: Survey evidence from Liberia and Sierra Leone. *SSM-Population Health, 10*, 100472.

Hewlett, B. S., & Amola, R. P. (2003). Cultural contexts of Ebola in northern Uganda. *Emerging Infectious Diseases, 9*, 1242–1248.

James, P. B., Wardle, J., Steel, A., & Adams, J. (2020). An assessment of Ebola-related stigma and its association with informal healthcare utilisation among Ebola survivors in Sierra Leone: A cross-sectional study. *BMC Public Health, 20*(182).

Konkor, I., Lawson, E. S., Antabe, R., McIntosh, M. D., Husbands, W., Wong, J., & Luginaah, I. (2020). An intersectional approach to HIV vulnerabilities and testing among heterosexual African Caribbean and black men in London, Ontario: Results from the weSpeak study. *Journal of Racial and Ethnic Health Disparities*, 1–10.

Lasuta, J. (2014). *Ebola victims face stigma in West Africa. 2014*. Available from: http://www.voanews.com/content/ebola-victimsface-stigma-in-west-africa/1902587.html. Accessed on 12 Jan 2022.

Lotsch, F., Schnyder, J., Goorhuis, A., & Grobusch, M. P. (2017). Neuropsychological long-term sequelae of Ebola virus disease survivors – A systematic review. *Travel Medicine and Infectious Disease, 18*, 18–23.

Luginaah, I. N., Kangmennaang, J., Fallah, M., Dahn, B., Kateh, F., & Nyenswah, T. (2016). Timing and utilization of antenatal care services in Liberia: Understanding the pre-Ebola epidemic context. *Social Science & Medicine, 160*, 75–86.

Obilade, T. T. (2015). Ebola virus disease stigmatization; the role of societal attributes. *International Archives of Medicine, 8*.

Overholt, L., Wohl, D. A., Fischer, W. A., 2nd, Westreich, D., Tozay, S., Reeves, E., Pewu, K., Adjasso, D., Hoover, D., Merenbloom, C., Johnson, H., Williams, G., Conneh, T., Diggs, J., Buller, A., McMillian, D., Hawks, D., Dube, K., & Brown, J. (2018). Stigma and Ebola survivorship in Liberia: Results from a longitudinal cohort study. *PLoS One, 13*(11), e0206595.

Rabelo, I., Lee, V., Fallah, M. P., Massaquoi, M., Evlampidou, I., Crestani, R., Decroo, T., Van den Bergh, R., & Severy, N. (2016). Psychological distress among Ebola survivors discharged from an Ebola treatment unit in Monrovia, Liberia – A qualitative study. *Frontiers in Public Health, 4*, 142.

Reid, A. E., Dovidio, J. F., Ballester, E., & Johnson, B. T. (2014). HIV prevention interventions to reduce sexual risk for African Americans: The influence of community-level stigma and psychological processes. *Social Science & Medicine, 103*, 118–125.

Shoman, H., Karafillakis, E., & Rawaf, S. (2017). The link between the west African Ebola outbreak and health systems in Guinea, Liberia, and Sierra Leone: A systematic review. *Globalization and Health, 13*(1), 1.

Smith-Morris, C. (2017). Epidemiological placism in public health emergencies: Ebola in two Dallas neighborhoods. *Social Science & Medicine, 179*, 106–114.

Tenkorang, E. Y. (2017). Ebola-related stigma in Ghana: Individual and community level determinants. *Social Science & Medicine, 182*, 142–149.

Vetter, P., Kaiser, L., Schibler, M., Ciglenecki, I., & Bausch, D. G. (2016). Sequelae of ebola-virus disease: The emergency within the emergency. *The Lancet Infectious Diseases, 16*(6), e82–91.6.

Vorauer, J. D. (2006). An information search model of evaluative concerns in intergroup interaction. *Psychological Review, 113*(4), 862.

WHO. (2016). *Clinical care for survivors of Ebola virus disease*. Retrieved from: https://apps.who.int/iris/bitstream/handle/10665/204235/WHO_EVD_OHE_PED_16.1_eng.pdf;jsessionid=3A2B9E6BC517404B0F9624417152304B?sequence=1. Accessed on 06 Apr 2020.

WHO. (2016a). Ebola virus disease situation report.

World Bank. (2021). *Sierra Leone, total population*. Retrieved from https://data.worldbank.org/indicator/SP.POP.TOTL?locations=SL. Accessed on 12 Jan 2022.

World Health Organization. (2020). *Ebola virus disease*. Retrieved from https://www.who.int/news-room/fact-sheets/detail/ebola-virus-disease.

Chapter 12
Geospatial Analysis of Tungiasis Disease Transmission Risk in East Africa

Mark A. Deka and Niaz Morshed

Introduction

Throughout history, populations worldwide have been exposed to various parasitic diseases that have caused considerable health, economic, and societal burdens (Brooker, 2007; Jefferey & Pia, 2002). One such disease, tungiasis, is found throughout sub-Saharan Africa, South America, and the Caribbean. Tungiasis predominately affects resource-poor communities with limited healthcare services (Elson et al., 2017). The exclusive agent is the female flea (jigger flea; chigoe flea) from the genus Tunga (order: Siphonaptera, family: hectopsyllidae). *Tunga penetrans*, the smallest known flea in the world (< 1 mm), causes disease by burrowing into the host's epidermis and depositing eggs, primarily in the hands, knees, heels, soles, and interdigital spaces (Chen et al., 2011; Heukelbach & Ugbomoiko, 2011; Pampiglione et al., 2009).

Once the flea enters the skin, it increases in size, causing itching and inflammation (Feldmeier et al., 2003). Severe disease can lead to the development of fissures, disfigurement, amputation, immobilization, and chronic lymphedema, as well as secondary infections like septicemia, tetanus, and toxic shock syndrome (Elson et al., 2017; Kimotho et al., 2015; Nyangacha et al., 2017; Obengui, 1989; Tadele, 2017; Deka, 2020). Geographically, the disease is common in rural areas, urban slums, and villages on the periphery of riparian and lake systems (Franck et al., 2003). Socioeconomic factors like poverty, illiteracy, and low hygiene standards are contributing factors (Ehrenberg & Ault, 2005; Wafula et al., 2016). As a result,

M. A. Deka (✉)
Centers for Disease Control and Prevention, Atlanta, GA, USA
e-mail: pmu5@cdc.gov

N. Morshed
Texas Health and Human Services Commission, Austin, TX, USA

J. A. Braimah et al. (eds.), *Health Geography in Sub-Saharan Africa*, Global Perspectives on Health Geography, https://doi.org/10.1007/978-3-031-37565-1_12

tungiasis prevalence rates range from 15% to 40% in some communities (Pampiglione et al., 2009). Domestic and sylvatic animals serving as zoonotic reservoirs include cats, dogs, chickens, goats, pigs, rats, cattle, and monkeys (Cooper, 1967; Heukelbach et al., 2001).

Ecological niche models (ENM) are powerful tools for characterizing the geography of various disease vectors, which increases knowledge of relevant parasites-reservoirs and their potential geographic distributions (Costa et al., 2014). Ecological niche modeling, combined with geospatial techniques, can improve targeted control measures and enhance the cost-effectiveness of disease control programs (Brooker, 2007; Deka, 2022; Deka et al., 2022a, b). Thus, the primary objectives of this exploratory analysis were to: (1) develop an ecological niche model delineating suitability for tungiasis; (2) and map the subsequent disease transmission risk to humans.

Data and Methods

Occurrence Data Collection and Spatial Thinning

The acquisition of tungiasis human and animal data was completed through a literature review spanning a temporal period from 1895 to 2022 (Chelimo, 2013; Dassoni et al., 2014; Mørkve, 2013; Mwakanyamale et al., 2015; Proctor, 1970; Ruttoh et al., 2012; Sentongo & Wabinga, 2014; Tadele, 2017; Wafula et al., 2016; Waruguru et al., 2015; Wiese et al., 2015; Alfred, 2009; Gadisa & Jote, 2019; Girma et al., 2018; Grupper & Potasman, 2012; Heukelbach & Ugbomoiko, 2007; Jeffreys, 1952; Johnston, 1895; Karuga, 2013; Kitching, 1912; Luchetti et al., 2007; Mazigo et al., 2011; Mutebi et al., 2015; Mwai et al., 2018, 2022; Mwangi et al., 2015; Nájera et al., 2019; Nyangacha et al., 2019; Jorga et al., 2020; Ramaswamy, 2016; Walker et al., 2017; Wiese, 2020; Yosef et al., 2018; Zabron, 2017). Additionally, we supplemented these data with *T. penetrans* occurrence data from the Global Biodiversity Information Facility (GBIF) ($n = 1$) (GBIF, 2022). Finally, tungiasis occurrence locations ($n = 53$) were georeferenced with the guidance of Google Earth Pro (Google Earth, 2022) and ESRI ArcGIS 10.8.1 (2022) (ESRI, 2022). To account for potential spatial autocorrelation in our data, we used the SDM toolbox 2.0 (Brown et al., 2017) extension in ArcGIS 10.8.1 (ESRI. ArcGIS desktop: release 10.8.1. Environmental Systems Research Institute, CA, USA) to filter occurrence locations at a 20 km threshold spatially. As a result, the final dataset featured 45 occurrence locations.

Study Area

The spatial extent used during the niche modeling process directly influences modeling outcomes (Barve et al., 2011; Costa et al., 2014). In this study, models were designed with a delimited spatial extent or the hypothesized area (**M**) encompassing Ethiopia, Kenya, Uganda, Rwanda, Burundi, and Tanzania

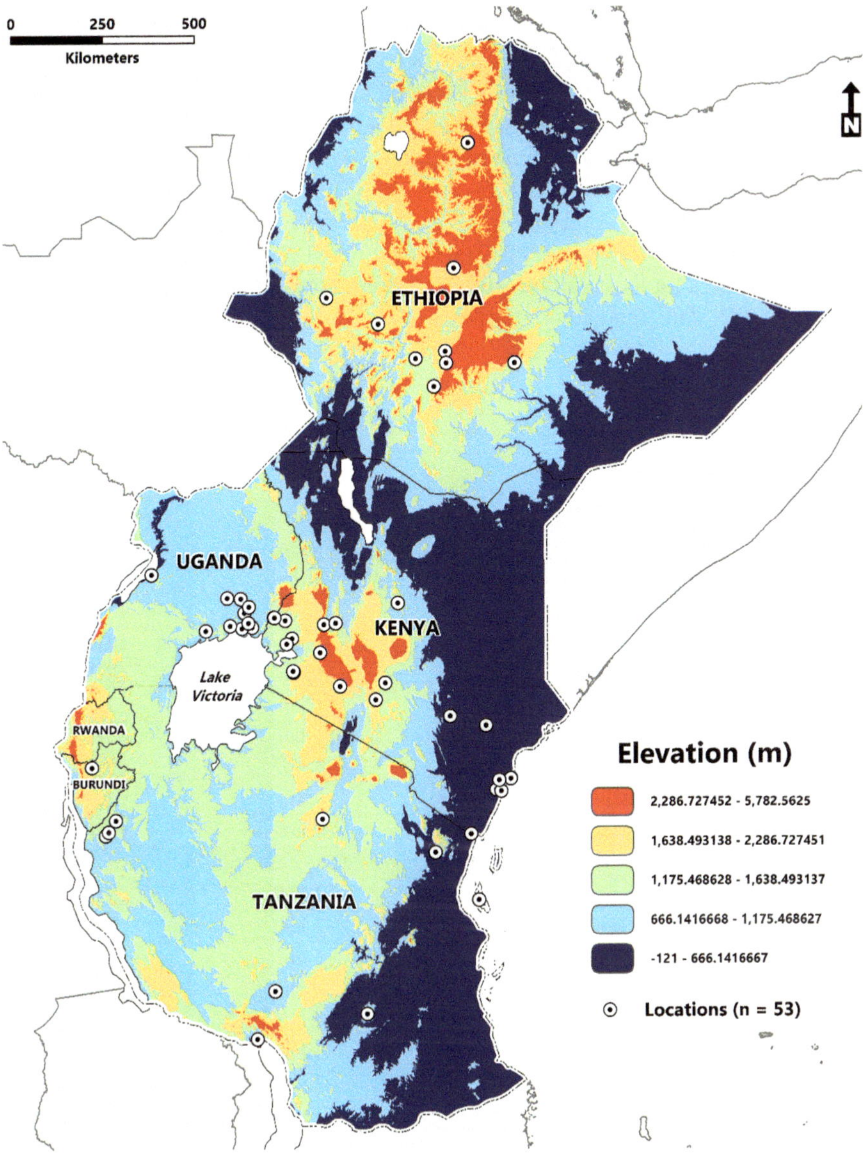

Fig. 12.1 Study area in relation to elevation (M region)

(Fig. 12.1). The estimation of **M** is based on the BAM (biotic, abiotic, movement) framework. In the BAM framework, the geographic distribution of a species is an intersection between biotic, abiotic, and movement factors (Soberón, 2005). The movement (**M**) aspect of the BAM framework is incorporated to understand the influences of spatial extent on modeling outcomes as well as the effects of model training, validation, and comparisons; while further considering the geographic accessibility (Barve et al., 2011).

Environmental Variables

We chose our environmental variables based on previous tungiasis modeling studies in Africa and Latin America (Deka, 2020; Deka & Heukelbach, 2022). Bioclimatic data (Table 12.1) representing temperature (bio1–11) and precipitation (bio12–19) were collected at a ~ 1 km resolution from the CHELSA v.2.1 (Chelsa Climate – Climatologies at high resolution for the earth's land surface areas (chelsa-climate.org)) (accessed on December 1, 2022) database (1981–2010) (Karger et al., 2017). Topographic data for the mean elevation and slope were obtained at a ~ 1 km resolution from the data repository, EarthEnv (Amatulli et al., 2018) (accessed on

Table 12.1 Predictor variables

Variables	Resolution	Source
Set1		
BIO1 – Annual mean temperature	~1 km	**CHELSA**
BIO2 – Mean diurnal range	~1 km	**CHELSA**
BIO3 – Isothermality	~1 km	**CHELSA**
BIO4 – Temperature seasonality	~1 km	**CHELSA**
BIO5 – Max temperature of warmest month	~1 km	**CHELSA**
BIO6 – Min temperature of coldest month	~1 km	**CHELSA**
BIO7 – Temperature annual range	~1 km	**CHELSA**
BIO8 – Mean temperature of wettest quarter	~1 km	**CHELSA**
BIO9 – Mean temperature of driest quarter	~1 km	**CHELSA**
BIO10 – Mean temperature of warmest quarter	~1 km	**CHELSA**
BIO11 – Mean temperature of coldest quarter	~1 km	**CHELSA**
Set 2		
BIO12 – Annual precipitation	~1 km	**CHELSA**
BIO13 – Precipitation of wettest month	~1 km	**CHELSA**
BIO14 – Precipitation of driest month	~1 km	**CHELSA**
BIO15 – Precipitation seasonality	~1 km	**CHELSA**
BIO16 – Precipitation of wettest quarter	~1 km	**CHELSA**
BIO17 – Precipitation of driest quarter	~1 km	**CHELSA**
BIO18 – Precipitation of warmest quarter	~1 km	**CHELSA**
BIO19 – Precipitation of coldest quarter	~1 km	**CHELSA**
Set3		
Sand	~1 km	SoilGrids
Silt	~1 km	SoilGrids
Clay	~1 km	SoilGrids
Soil pH	~1 km	SoilGrids
Set4		
Elevation	~1 km	**EarthEnv**
Slope	~1 km	**EarthEnv**
Set5		
Enhanced vegetation index (EVI)	~1 km	WorldGrids
Land surface temperature (LST)	~1 km	WorldGrids

December 1, 2022). We selected edaphic characteristics related to sand, clay, pH, and silt content (5 cm depth, g/100 (w%)) (~1 km resolution) from the World Soil Information database (International Soil Reference and Information Centre (ISRIC)) (ISRIC, 2018) (SoilGrids250m 2.0) (accessed on December 1, 2022). We further included vegetation and land surface temperature information with a moderate-resolution imaging spectroradiometer (MODIS) (National Aeronautics and Space Administration (NASA)) (monthly mean) enhanced vegetation index (EVI) and land-surface temperature (LST) coverages (~1 km). These data were obtained from the WorldGrids data archive (Hengl, 2018) (accessed on December 1, 2022).

Principal Component Analysis (PCA)

A principal component analysis (PCA) reduced the levels of multicollinearity between the environmental variables. This method has been applied previously in ecological niche modeling studies (Deka et al., 2022a, b; Romero-Alvarez et al., 2020). The covariates were separated into five sets to calculate the leading principal components (PCs) with software package, Niche Analyst (Qiao et al., 2016; http://nichea.sourceforge.net/). For the **CHELSA** data, the first five PCs were retained for temperature (99%) (BIO1-11) and the first three for precipitation (99%) (BIO12-19). In addition, three PCs for soil were retained (100%), as were the first two PCs for EVI and LST (100%) and elevation and slope (100%).

Ecological Niche Modeling

To model the ecological niche of tungiasis, we used the maximum entropy algorithm (Maxent) (Phillips et al., 2006) found in the R package (R Studio, 2022), 'Maxnet' v.0.1.4.24 (Philips et al., 2018). Prior to modeling, we tuned our data by varying the level of 'regularization' to the optimal level of model complexity with the R package (R Studio team PBC, 2022) 'ENMeval' (Kass et al., 2022), integrated with the ENM application, 'Wallace' (Kass et al., 2018). We set a range for our regularization multipliers (RM) from 0.5 to 5.0 at increments of 0.5 (0.5, 1, 1.5, 2, 2.5, 3, 3.5, 4, 4.5, 5) with five different feature class combinations (L, LQ, H, LQH, LQHP; L = linear, Q = Quadratic, H = hinge, and P = product) (50 candidate models) and specified a 'chekerboard2' spatial partition method (Deka & Heukelbach, 2022). The best candidate model settings were based on the lowest (< 0) corrected Akaike information criterion (AICc) (delta. AICc) value (Burnham & Anderson, 2004; Deka & Heukelbach, 2022).

We validated the predictive ability of our model with the partial ROC (pROC) (Peterson et al., 2008) metric (5% training omission error rate, i.e., sensitivity >0.95). When compared to the traditional ROC metric (Elith et al., 2011), the pROC accesses omission errors for independent points and the predicted suitable area.

Partial AUC values are derived from the ratio between the AUC and the null expectations of the AUC. The corresponding pROC ratios thus range from 0 to 2 (values of 1 = random performance; Peterson, 2012; Peterson et al., 2008). Partial AUC ratios greater than 1 and closer to 2 indicate better predictive potential. The pROC tests were conducted with the R (Team R, 2022) package, 'ENMGadgets' (Barve & Barve, 2013).

Estimating Human Disease Transmission Risk

Human disease transmission was mapped by linking the ecological niche of tungiasis with human population density. Two primary factors determined these risk zones: (i) the ecological niche of tungiasis and (ii) human population density (Alaniz et al., 2017; Carvajal et al., 2020; Deka et al., 2022a, b; Deka, 2022). This geospatial method has foundations in precision public health, which integrates geolocated health data and maps to identify regions of elevated disease transmission risk (Osgood-Zimmerman et al., 2018; Reich & Haran, 2018; Deka, 2022). First, we obtained human population density data (per pixel cell) (~1 km) (2020) for Ethiopia, Kenya, Uganda, Burundi, Rwanda, and Tanzania from the WorldPop project (Open Spatial Demographic Data and Research – WorldPop) (accessed on December 1, 2022). Then, we first reclassified the data into four categories: null (< 1 human/km2), low (> 1–10 humans/km2), medium (> 11–100 humans/km2), and high (> 101 humans/km2).

For each classification, we assigned new values of null = 0, low = 1, medium = 2, high = 3. Second, the niche model was then reclassified into four categories: null (0), low (1), medium (2), and high (3), with an equal interval classification type. Following reclassification, both datasets were multiplied using the Raster Calculator tool ArcGIS 10.8.1 (ESRI, 2022) (ArcGIS desktop: release 10.8.1. Environmental Systems Research Institute; www.esri.com). In addition, we spatially quantified the estimated total population living within these designated risk zones (very high, high, medium, low, very low) with 2020 population count (per pixel cell) (~1 km) data for Ethiopia, Kenya, Uganda, Burundi, Rwanda, and Tanzania. These data were also obtained from the WorldPop project (Open Spatial Demographic Data and Research – WorldPop) (accessed on December 1, 2022).

Results

We compiled a database of tungiasis human and animal occurrence data (n = 53) from 1895 to 2022 (Burundi = 1; Ethiopia = 9; Kenya = 24; Tanzania = 9; Uganda = 10). These data were located at an average elevation of 1231 meters (min – 24 meters, max – 2523 meters). The predicted ecological niche (Fig. 12.2) includes the Ethiopian Highlands, Uganda, western (Western, Nyanza, Rift Valley), central (Central and Rift Valley), and coastal Kenya (Coast), the African Great Lakes, Rwanda, Burundi, northeastern (Mara, Arusha, Kilimanjaro, Manyara), southern (Ruvuma, Njombe, Morogoro), coastal (Tanga, Pwani, Lindi, Mtwara), and western (Kigoma, Katavi, Rukwa, Mbeya) Tanzania. The predicted ecological niche is most influenced by precipitation (Set2) (PC1–28.1%), soil (Set3) (PC3–25.8%, PC2–5.9%), and temperature (Set1) (PC2–6.2%). The model was influenced to a lesser degree by vegetation (EVI) – land surface temperature (LST) (Set5) (PC2–5%), and topography (Set4) (PC2–4.2%, PC1–2.9%) (Table 12.2). The ideal maxent model settings based on the creation of 50 candidate models were linear (L), quadratic (Q) features, and a regularization multiplier = 2.5 (delta. AICc = 0). The partial ROC (pROC) test determined that the model had a high degree of predictive accuracy and performed better than random based on a ratio of observed to null expectations value of 1.68 (min – 1.31; max – 1.91).

The disease transmission risk to humans (Fig. 12.3) is greatest in Burundi, Rwanda, southern, central, and eastern Uganda, as well as central and western Kenya. Risk in Tanzania is concentrated on the southern shores of Lake Victoria, Lake Tanganyika, and Lake Malawi. Inland risk regions are also found in southern Tanzania, stretching from the Tanga, Zanzibar, and Dar es Salaam in the east, to the west, through Morogoro, Ruvuma, and Njombe. In addition, medium-high risk is concentrated along the coast of the Indian Ocean from Lamu (Kenya) and Mombasa (Kenya) in the north to Pwani (Tanzania) and Lindi (Tanzania) in the south. Within Ethiopia, disease transmission risk is greatest in Amhara, Oromiya, Benishangul-Gumuz, Southern Nations, Nationalities, People's Region, and Addis Ababa. The estimated population living in the (1) very high zone is 109,613,668, 79,234,296 reside in the (2) high-risk zone, while an additional 22,995,266 live in the (3) medium-risk zone. Finally, an estimated 2,361,921 reside within the (4) low-risk zone and 40,389,059 within the (5) very low zone, respectively (Table 12.3).

Discussion

This study mapped the ecological niche of tungiasis in East Africa and the disease transmission risk to human populations using ecological modeling and precision mapping analysis. Our model predicted (Fig. 12.2) ecological suitability for tungiasis throughout Burundi, Rwanda, the Ethiopian Highlands, southern, central, and eastern Uganda, western and central Kenya, and much of northern, western, and southern Tanzania. Suitability along coastal regions is concentrated in a north-south band stretching from Lamu (Kenya) and Mombasa (Kenya) in the north to Pwani (Tanzania), Lindi (Tanzania), and Mtwara (Tanzania). The ecological niche is most influenced by precipitation (Set2) (PC1–28.1%), soil (Set3) (PC3–25.8%, PC2–5.9%), and temperature (Set1) (PC2–6.2%).

Model predictive accuracy according to the partial ROC (pROC) test was strong, with an average ratio of observed to the null expected value of 1.68 (min – 1.31;

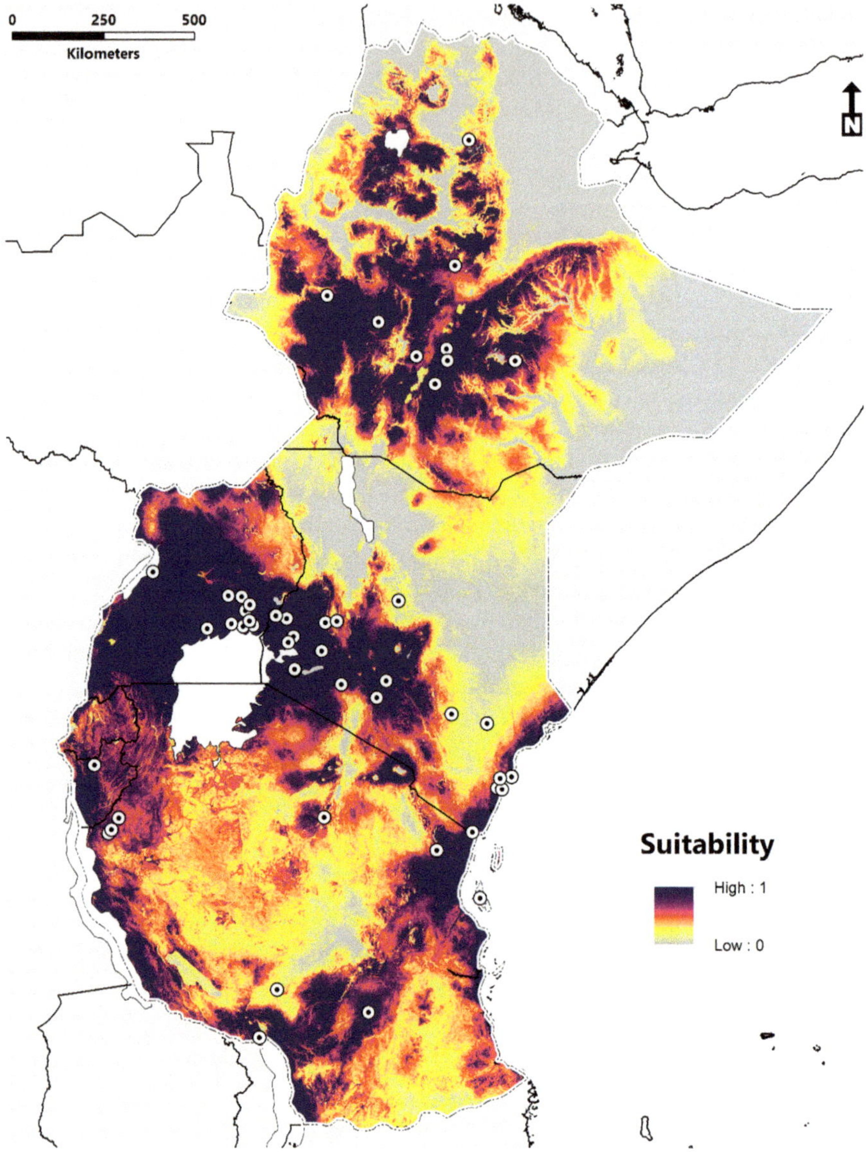

Fig. 12.2 Ecological niche of tungiasis in East Africa. Best model – Maxent (FC = LQ, RM – 2.5). Occurrence locations (*n* = 53) are visualized as white-black dots

max – 1.91). The highest disease transmission risk (Fig. 12.3) is concentrated in Burundi, Rwanda, Uganda, the Ethiopian Highlands, western and central Kenya, and southern, western, and coastal regions of Tanzania. In total, an estimated 109,613,668 people reside in the (1) very high-risk zone, as well as an additional 79,234,296 in the (2) high-risk zone. Additionally, 22,995,266 reside in the (3) medium-risk zone, 2,361,921 reside within the (4) low-risk zone, and an estimated

Table 12.2 Variable contributions

Variable	Model contribution
Set2 – Precipitation PC1	28.1
Set3 – Soil PC3	25.8
Set1 – Temperature PC2	6.2
Set3 – Soil PC2	5.9
Set5 – EVI/LST PC2	5
Set1 – Temperature PC1	4.48
Set1 – Temperature PC5	4.4
Set4 – Physical PC2	4.2
Set2 – Precipitation PC3	3.9
Set4 – Physical PC1	2.9

40,389,059 reside in the (5) very low zone, respectively. For the first time, this research mapped the disease transmission risk associated with tungiasis through the linkage of ecological niche modeling and human population density.

Ecological variables of importance to the model were precipitation (bio12–19), soil conditions (sand, silt, clay, soil pH), and temperature (bio1–11). Previous studies have identified low precipitation and high temperature as factors in the increased prevalence of tungiasis (Mwangi et al., 2015; Heukelbach et al., 2005). It has been hypothesized that heavy rain events can disrupt the superficial soil layers and wash away the off-host ectoparasites (Linardi et al., 2010). Similarly, soil factors have been recognized as having a positive correlation to the prevalence of tungiasis. For example, research in Murang'a County, Kenya, found that factors like soil pH and moisture levels influenced the prevalence of tungiasis by up to 33% and *T. penetrans* populations by 39.7% (Mwangi et al., 2015). In addition, neutral and alkaline soil pH values between 6.6 and 7.8 significantly affect *T. penetrans* development and thus are more favorable than acidic soil types (Ngunjiri & Keiyoro, 2011).

This research applied geographic information science (GIS) to estimate the ecological niche of tungiasis and the risk to human populations in East Africa. Ecological niche modeling has been applied for a variety of epidemiological purposes, such as the geographic estimation of filovirus (***Filoviridae***) emergence (Peterson et al., 2004) and avian influenza (Williams & Peterson, 2009). These high-resolution maps can provide policymakers and local public health organizations with detailed information on areas where intervention campaigns can be most successful in lessening the health burden associated with the disease.

Neglected Tropical Disease's (NTDs) like tungiasis are amplified by poverty, inadequate or non-existent sanitation and waste disposal services, unpaved streets, mud floors in houses, and infestations of mice and rats (Heukelbach et al., 2001). In addition, poverty can enhance zoonotic transmission through the increased presence of stray dogs, cats, pigs, and rodents (Pampiglione et al., 2009; Pilger et al., 2008). For example, in Kenya, Africa Health and Development International (AHADI) reported 2.7 million cases of tungiasis in 2009, with an estimated 25% of Kenyan children at risk of suffering from the disease (Kenya Ministry of Health, 2014; Ngunjiri & Keiyoro, 2011). Likewise, in southeastern and northeastern Uganda, the

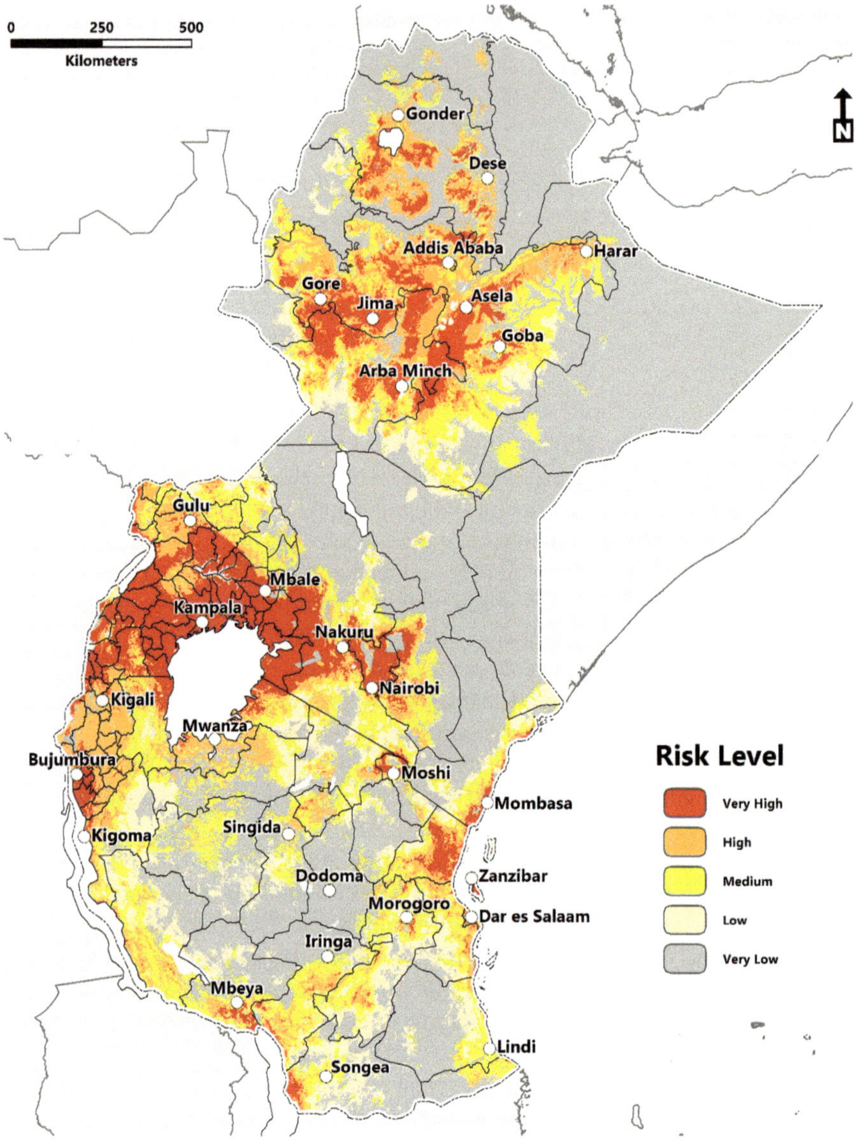

Fig. 12.3 Tungiasis human disease transmission risk in East Africa

prevalence of tungiasis is high; these areas are identified as some of the poorest in the country (Kenya Ministry of Health, 2014; Mutebi et al., 2017).

This study predicted that much of Uganda is suitable for tungiasis ecologically. The risk to human populations is equally as widespread geographically (Fig. 12.4). High-risk zones encompass much of southern, central, eastern, and northwest Uganda. These include population centers like Mbarara, Masaka, Kampala, Gulu,

Table 12.3 The estimated population (2020) within the designated risk zones

Variable	Population
1. Very high	109,613,668
2. High	79,234,296
3. Medium	22,995,266
4. Low	2,361,921
5. Very low	49,389,059

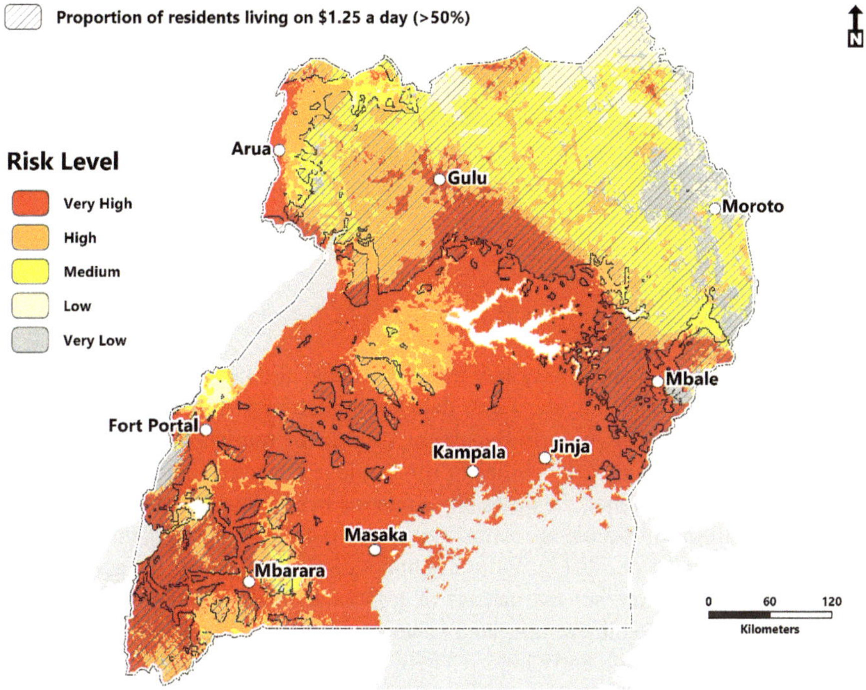

Fig. 12.4 Tungiasis human disease transmission risk in Uganda in relation to areas where greater than 50% of the population lives on $1.25 a day

and Arua. When compared to the distribution of the population living on $1.25 a day (WorldPop project) (Open Spatial Demographic Data and Research – WorldPop) (accessed on December 1, 2022), spatial overlap between these risk zones is prevalent in the south near Mbarara, central, north (Gulu and Arua), and the eastern region, in areas surrounding Mbale.

In addition, the threat of climate change could increase the prevalence of tungiasis in high-risk areas throughout East Africa. Because of the seasonal variation in tungiasis, climate change could alter transmission dynamics and increase the severity of outbreaks (Deka & Heukelbach, 2022). The effects of climate change are cited as a critical factor, for example, in the migration of vector-borne diseases into new agro-climatic zones (Biswas, 2022). This includes diseases like Dengue fever, West Nile virus, Japanese encephalitis, and Chikungunya (Biswas, 2022). Climate change will likely cause significant impacts on low-resource populations, which are often most impacted by NTDs (Deka & Heukelbach, 2022; Bryson et al., 2020).

It is important to note that ecological suitability does not necessarily imply presence. As a disease with a focal distribution, many factors are related to disease transmission risk and endemicity; these include cultural habits (i.e., walking barefoot, sitting/resting on the ground), the presence of slums/resource-poor communities, free-roaming animals, and urban vs rural populations (Deka & Heukelbach, 2022). The disease transmission risk maps should be viewed similarly, as this model is linked to population density. Regions with low population density will often be highlighted as low risk for tungiasis. However, this is not always correct, as resource-poor communities in rural areas often face increased challenges in preventing the disease (Heukelbach et al., 2001; Feldmeier et al., 2003). East Africa is cited as being the most isolated region in Africa, with an estimated 74.4% of the population living in rural areas (Deka, 2020; Amzat & Razum, 2017).

Study Limitations

We acknowledge that this study has methodological limitations that impede the potential scope of our analysis. First, data on human and animal occurrences is limited. Most published reports document disease in tourists returning from a vacation abroad or local scale studies in rural village enclaves (Pampiglione et al., 2009). Minimal occurrence data presents a serious public health issue, considering that tens of millions of people are estimated to be at risk of Tungiasis in the Americas and Africa (Elson et al., 2017). Future collaborations are needed to facilitate the development of a comprehensive and active geospatial database linking information from satellite-derived soil, climate, land cover data, and medical intelligence. These collaborations can significantly improve geographically targeted strategies and our understanding of Tungiasis disease ecology and estimates of spatial heterogeneity.

Conclusion

In summary, our study provides one of the few geographic descriptions of tungiasis in East Africa. This research provides an evidence-based approach that can be applied to enhance population-based studies through the application and utilization of geospatial information and analysis. Even with the above-stated limitations, this study provides a roadmap for future research concerning the geographic distribution of tungiasis and its risk to humans. Hopefully, this study will represent another chapter in the international drive toward greater awareness of tungiasis.

Acknowledgments This project was supported in part by an appointment to the Research Participation Program at the Centers for Disease Control and Prevention administered by the Oak Ridge Institute for Science and Education through an interagency agreement between the US Department of Energy and the Centers for Disease Control and Prevention.

References

Alaniz, A. J., Bacigalupo, A., & Cattan, P. E. (2017). Spatial quantification of the world population potentially exposed to Zika virus. *International Journal of Epidemiology, 46*(3), 966–975.

Alfred, K. (2009). Factors Associated with Jigger Infestation in Kituro, Baringo Central District. *A Rapid Appraisal Report*

Amatulli, G., Domisch, S., Tuanmu, M. N., Parmentier, B., Ranipeta, A., Malczyk, J., & Jetz, W. (2018). A suite of global, cross-scale topographic variables for environmental and biodiversity modeling. *Scientific Data, 5*(1), 1–15.

Amzat, J., & Razum, O. (2017). *Towards a sociology of health discourse in Africa*. Springer.

Barve, N., & Barve, V. (2013). *ENMGadgets: Tools for pre and post-processing in ENM workflows*. https://Github.Com/Vijaybarve/ENMGadgets

Barve, N., Barve, V., Jiménez-Valverde, A., Lira-Noriega, A., Maher, S. P., Peterson, A. T., et al. (2011). The crucial role of the accessible area in ecological niche modeling and species distribution modeling. *Ecological Modelling, 222*(11), 1810–1819. https://doi.org/10.1016/j.ecolmodel.2011.02.011

Biswas, B. K. (2022). Effect of climate change on vector-borne disease. In *Emerging issues in climate smart livestock production* (pp. 263–316). Academic Press.

Brooker, S. (2007). Spatial epidemiology of human schistosomiasis in Africa: Risk models, transmission dynamics, and control. *Transactions of the Royal Society of Tropical Medicine and Hygiene, 101*(1), 1–8. https://doi.org/10.1016/j.trstmh.2006.08.004

Brown, J. L., Bennett, J. R., & French, C. M. (2017). SDMtoolbox 2.0: The next-generation python-based GIS toolkit for landscape genetic, biogeographic, and species distribution model analyses. *PeerJ, 5*, e4095.

Bryson, J. M., Bishop-Williams, K. E., Berrang-Ford, L., Nunez, E. C., Lwasa, S., Namanya, D. B., et al. (2020). Neglected tropical diseases in the context of climate change in East Africa: A systematic scoping review. *The American Journal of Tropical Medicine and Hygiene, 102*(6), 1443.

Burnham, K. P., & Anderson, D. R. (2004). Multimodel inference: Understanding AIC and BIC in model selection. *Sociological Methods & Research, 33*(2), 261–304.

Carvajal, J. G., Alaniz, A. J., Carvajal, M. A., Acheson, E. S., Cruz, R., Vergara, P. M., & Cogliati, M. (2020). Expansion of the emerging fungal pathogen cryptococcus bacillisporus into America: Linking phylogenetic origin, geographical spread, and population under exposure risk. *Frontiers in Microbiology, 11*, 2117.

Chelimo, J. J. (2013). *Risk factors associated with jigger infestation in kitany location, Keiyo Marakwet County*. Moi University.

Chen, C. W., Thong, H. Y., & Jee, S. H. (2011). Tungiasis: A case report and review of the literature. *Dermatologica Sinica, 29*(1), 29–31. https://doi.org/10.1016/j.dsi.2011.01.004

Cooper, J. E. (1967). An outbreak of Tunga penetrans in a pig herd. *The Veterinary Record, 80*, 365–366. https://doi.org/10.1136/vr.80.11.365

Costa, J., Dornak, L. L., Almeida, C. E., & Peterson, A. T. (2014). Distributional potential of the Triatoma brasiliensis species complex at present and under scenarios of future climate conditions. *Parasites & Vectors, 7*(1), 1–10.

Dassoni, F., Polloni, I., Margwe, S. B., & Veraldi, S. (2014). Tungiasis in northern Tanzania: A clinical report from Qameyu village, Babati District, Manyara region. *Journal of Infection in Developing Countries, 8*(11), 1456–1460. https://doi.org/10.3855/jidc.4323

Deka, M. A. (2020). Mapping the geographic distribution of tungiasis in sub-Saharan Africa. *Tropical Medicine and Infectious Disease, 5*(3), 122.

Deka, M. A. (2022). Predictive risk mapping of schistosomiasis in Madagascar using ecological niche modeling and precision mapping. *Tropical Medicine and Infectious Disease, 7*(2), 15.

Deka, M. A., & Heukelbach, J. (2022). Distribution of Tungiasis in Latin America: Identification of areas for potential disease transmission using an ecological niche model. *The Lancet Regional Health-Americas, 5*, 100080.

Deka, M. A., Marston, C. K., Garcia-Diaz, J., Drumgoole, R., & Traxler, R. M. (2022a). Ecological niche model of Bacillus cereus group isolates containing a homologue of the pXO1 anthrax toxin genes infecting metalworkers in the United States. *Pathogens, 11*(4), 470.

Deka, M. A., Vieira, A. R., & Bower, W. A. (2022b). Modeling the ecological niche of naturally occurring anthrax at global and circumpolar extents using an ensemble modeling framework. *Transboundary and Emerging Diseases., 69*, e2563.

Ehrenberg, J. P., & Ault, S. K. (2005). Neglected diseases of neglected populations: Thinking to reshape the determinants of health in Latin America and the Caribbean. *BMC Public Health, 5*, 1–13. https://doi.org/10.1186/1471-2458-5-119

Elith, J., Phillips, S. J., Hastie, T., Dudík, M., Chee, Y. E., & Yates, C. J. (2011). A statistical explanation of MaxEnt for ecologists. *Diversity and Distributions, 17*(1), 43–57.

Elson, L., Wright, K., Swift, J., & Feldmeier, H. (2017). Control of Tungiasis in absence of a roadmap: Grassroots and global approaches. *Tropical Medicine and Infectious Disease, 2*(3), 33. https://doi.org/10.3390/tropicalmed2030033

ESRI. (2022). *Environmental System Research Institute (ESRI): ArcGIS Desktop (Version 10.8.1)* [Software]. Retrieved from https://www.esri.com/en-us/home

Feldmeier, H., Eisele, M., Sabóia-Moura, R. C., & Heukelbach, J. (2003). Severe tungiasis in underprivileged communities: Case series from Brazil. *Emerging Infectious Diseases, 9*(8), 949–955. https://doi.org/10.3201/eid0908.030041

Franck, S., Feldmeier, H., & Heukelbach, J. (2003). Tungiasis: More than an exotic nuisance. *Travel Medicine and Infectious Disease, 1*(3), 159–166. https://doi.org/10.1016/j.tmaid.2003.09.005

Gadisa, E., & Jote, K. (2019). Prevalence and factors associated with intestinal parasitic infection among under-five children in and around Haro Dumal Town, Bale Zone, Ethiopia. *BMC Pediatrics, 19*(1), 1–8.

GBIF.org. (2022, Decemer 1). *GBIF occurrence download.* https://doi.org/10.15468/dl.xcpprz.

Girma, M., Astatkie, A., & Asnake, S. (2018). Prevalence and risk factors of tungiasis among children of Wensho district, Southern Ethiopia. *BMC Infectious Diseases, 18*(1), 1–7.

Google Earth. (2022). *The world's most detailed globe*. Retrieved December 1, 2022, from https://www.google.com/earth/

Grupper, M., & Potasman, I. (2012). Outbreak of tungiasis following a trip to Ethiopia. *Travel Medicine and Infectious Disease, 10*(5–6), 220–223.

Hengl, T. (2018). *WorldGrids archived layers at 1 km to 20 km spatial resolution*. https://zenodo.org/record/1637816. Accessed 1 Dec 2022.

Heukelbach, J., & Ugbomoiko, U. S. (2007). Tungiasis in the past and present: A dire need for intervention. *Nigerian Journal of Parasitology, 28*(1), 1–5.

Heukelbach, J., & Ugbomoiko, U. S. (2011). Editorial: Tungiasis in the past and present: A dire need for intervention. *Nigerian Journal of Parasitology, 28*(1), 1–5. https://doi.org/10.4314/njpar.v28i1.37849

Heukelbach, J., Sales De Oliveira, F. A., Hesse, G., & Feldmeier, H. (2001). Tungiasis: A neglected health problem of poor communities. *Tropical Medicine and International Health, 6*(4), 267–272. https://doi.org/10.1046/j.1365-3156.2001.00716.x

Heukelbach, J., Wilcke, T., Harms, G., & Feldmeier, H. (2005). Seasonal variation of tungiasis in an endemic community. *American Journal of Tropical Medicine and Hygiene, 72*, 145–149.

ISRIC. (2018). *International Soil Reference and Information Centre (ISRIC)*. Retrieved October 9, 2018, from https://www.isric.org/

Jefferey, S., & Pia, M. (2002). The economic and social burden of malaria. *Nature, 415*(6872), 680–685. https://doi.org/10.1038/415680a

Jeffreys, M. D. W. (1952). Pulex Penetrans the Jigger's arrival and spread in Africa. *South African Journal of Science, 48*(8), 249–255.

Johnston, H. (1895). *The river Congo, from its mouth to Bolobo: A general description of the natural history and anthropology of its Western Basin*. Sampson Low, Marston & Company.

Jorga, S. D., Lulu Dessie, Y., Reshad Kedir, M., & Donacho, D. O. (2020). Prevalence of Tungiasis and its risk factors of among children of Mettu woreda, Southwest Ethiopia, 2020. *PLoS One, 17*(1), e0262168.

Karger, D. N., Conrad, O., Böhner, J., Kawohl, T., Kreft, H., Soria-Auza, R. W., et al. (2017). Climatologies at high resolution for the earth's land surface areas. *Scientific Data, 4*(1), 1–20.

Karuga, J. M. (2013). *Factors contributing to prevalence of jigger infestation among community members of Mugumoini sublocation*. Gatanga District Kenya Kenya Medical Training College.

Kass, J. M., Vilela, B., Aiello-Lammens, M. E., Muscarella, R., Merow, C., & Anderson, R. P. (2018). Wallace: A flexible platform for reproducible modeling of species niches and distributions built for community expansion. *Methods in Ecology and Evolution, 9*(4), 1151–1156.

Kass, J. M., Muscarella, R., Galante, P. J., Bohl, C. L., Pinilla-Buitrago, G. E., Boria, R. A., et al. (2022). ENMeval 2.0: Redesigned for customizable and reproducible modeling of species' niches and distributions. *Methods in Ecology and Evolution, 12*(9), 1602–1608.

Kenya Ministry of Health. (2014). *National policy guidelines on prevention and control of jigger infestations division of environmental health*. Retrieved from http://www.jigger-ahadi.org/National%20Policy%20Guidelines%20for%20Prevention%20and%20Control.pdf

Kimotho, S., Miller, A. N., & Ngure, P. (2015). Managing communication surrounding tungiasis stigma in Kenya. *Communication, 41*(4), 523–542. https://doi.org/10.1080/02500167.2015.1100646

Kitching, A. L. (1912). *On the backwaters of the Nile: Studies of some child races of Central Africa*. C. Scribner's Sons.

Linardi, P. M., Calheiros, C. M. L., Campelo-Junior, E. B., Duarte, E. M., Heukelbach, J., & Feldmeier, H. (2010). Occurrence of the off-host life stages of Tunga penetrans (Siphonaptera) in various environments in Brazil. *Annals of Tropical Medicine & Parasitology, 104*(4), 337–345.

Luchetti, A., Trentini, M., Pampiglione, S., Fioravanti, M. L., & Mantovani, B. (2007). Genetic variability of Tunga penetrans (Siphonaptera, Tungidae) sand fleas across South America and Africa. *Parasitology Research, 100*(3), 593–598.

Mazigo, H. D., Bahemana, E., Dyegura, O., Mnyone, L. L., Kweka, E. J., Zinga, M., et al. (2011). Severe tungiasis in Western Tanzania: Case series. *Journal of Public Health in Africa, 2*(2), 21.

Mørkve, Å. (2013). *Getting rid of the plague: Jiggers removal program in Bungoma, Kenya. Community and health workers perspectives on tungiasis in a high prevalence area* (Master's thesis). The University of Bergen.

Mutebi, F., Krücken, J., Feldmeier, H., Waiswa, C., Mencke, N., Sentongo, E., & von Samson-Himmelstjerna, G. (2015). Animal reservoirs of zoonotic tungiasis in endemic rural villages of Uganda. *PLoS Neglected Tropical Diseases, 9*(10), e0004126.

Mutebi, F., Eneku, W., Waiswa, C., Feldmeier, H., von Samson-Himmelstjerna, G., Krücken, J., et al. (2017). Animal and human tungiasis-related knowledge and treatment practices among animal keeping households in Bugiri District, south-eastern Uganda. *Acta Tropica, 177*(October 2017), 81–88. https://doi.org/10.1016/j.actatropica.2017.10.003

Mwai, J., Mutai, J., Karanja, S., & Karani, G. (2018). Factors influencing practices towards water, sanitation and hygiene with occurrence of Tungiasis among pupils in schools with a feeding Programme in Ganze Sub County, Kenya. *Global Journal of Health Sciences, 3*(1), 1–23.

Mwai, J., Nyole, D., Abdia, M. H., & Omogi, J. (2022). Factors associated with Tungiasis among school-age children in Kwale count, rural Kenya. *International Health, 15*(1), 85–92.

Mwakanyamale, J. G., Towett, R. K., Mtango, F., Bundala, J., & Kisanga, F. (2015). Contributions of socio-economic and cultural factors in Tungiasis at Kwakombo village in Korogwe district, Tanzania. *IMTU Medical Journal, 6*, 54–61.

Mwangi, J. N., Ozwara, H. S., Motiso, J. M., & Gicheru, M. M. (2015). Characterization of Tunga penetrans antigens in selected epidemic areas in Murang'a county in Kenya. *PLoS Neglected Tropical Diseases, 9*. https://doi.org/10.1371/journal.pntd.0003517

Nájera, S. M., Santisteban, G. N., & A. (2019). Tungiasis: A highly neglected disease among neglected diseases. Case series from Nduta refugee camp (Tanzania). *Oxford Medical Case Reports, 2019*(6), omz049.

Ngunjiri, J., & Keiyoro, P. (2011). *Soil factors influencing the occurrence of Tunga penetrans in Kenya* (vol. 7, no. 30, pp. 53–54). Germany: Lap Lambert Academic Publishing GmbH & Company KG.

Nyangacha, R. M., Odongo, D., Oyieke, F., Ochwoto, M., Korir, R., Ngetich, R. K., et al. (2017). Secondary bacterial infections and antibiotic resistance among tungiasis patients in Western, Kenya. *PLoS Neglected Tropical Diseases, 11*(9), 1–11. https://doi.org/10.1371/journal.pntd.0005901

Nyangacha, R. M., Odongo, D., Oyieke, F., Bii, C., Muniu, E., Chasia, S., & Ochwoto, M. (2019). Spatial distribution, prevalence and potential risk factors of Tungiasis in Vihiga County, Kenya. *PLoS Neglected Tropical Diseases, 13*(3), e0007244.

Obengui. (1989). Tungiasis and tetanus at the University Hospital Center in Brazzaville. *La Tungose et Le Tetanos Au C.H.U. de Brazzaville.*

Osgood-Zimmerman, A., Millear, A. I., Stubbs, R. W., Shields, C., Pickering, B. V., Earl, L., Graetz, N., Kinyoki, D. K., Ray, S. E., & Bhatt, S. (2018). Mapping child growth failure in Africa between 2000 and 2015. *Nature, 555*(7694), 41–47.

Pampiglione, S., Fioravanti, M. L., Gustinelli, A., Onore, G., Mantovani, B., Luchetti, A., & Trentini, M. (2009). Sand flea (Tunga spp.) infections in humans and domestic animals: State of the art. *Medical and Veterinary Entomology, 23*(3), 172–186. https://doi.org/10.1111/j.1365-2915.2009.00807.x

Peterson, A. T. (2012). Niche modeling: Model evaluation. *Biodiversity Informatics, 8*(1), 4300. 10.17161/bi.v8i1.4300.

Peterson, A. T., Bauer, J. T., & Mills, J. N. (2004). Ecologic and geographic distribution of filovirus disease. *Emerging Infectious Diseases., 10*, 40.

Peterson, A. T., Papes, M., & Soberón, J. (2008). Rethinking receiver operating characteristic analysis applications in ecological niche modeling. *Ecological Modelling, 213*(1), 63–72.

Philips, S. J., Dudík, M., & Schapire, R. E. (2018). *MaxEnt software for modeling species niches and distributions (version 3.4. 1)*. American Museum of Natural History.

Phillips, S. B., Anderson, R. P., & Schapire, R. E. (2006). Maximum entropy modeling of species geographic distributions. *International Journal of Global Environmental Issues, 190*(2006), 231–259. https://doi.org/10.1016/j.ecolmodel.2005.03.026

Pilger, D., Schwalfenberg, S., Heukelbach, J., Witt, L., Mencke, N., Khakban, A., & Feldmeier, H. (2008). Controlling tungiasis in an impoverished community: An intervention study. *PLoS Neglected Tropical Diseases, 2*(10), e324. https://doi.org/10.1371/journal.pntd.0000324

Proctor, E. M. (1970). Tunga penetrans acquired while traveling in Africa. *Canadian Journal of Infectious Diseases, 5*, 82–83. https://doi.org/10.1155/1994/781314

Qiao, H., Peterson, A. T., Campbell, L. P., Soberón, J., Ji, L., & Escobar, L. E. (2016). NicheA: Creating virtual species and ecological niches in multivariate environmental scenarios. *Ecography, 39*(8), 805–813.

R Core Team. (2022). *R: A language and environment for statistical computing (Version 3.2.4)* [Software]. Retrieved from https://cran.r-project.org/bin/windows/base/old/3.2.4/

Ramaswamy, V. (2016). Assessment of socioeconomic status and the prevalence of Tungiasis in Jimma and Wolaita Sodo, Ethiopia. *International Journal of Integrative Medical Sciences, 3*, 211–215.

Reich, B. J., & Haran, M. (2018). Precision maps for public health. *Nature, 555*(7694), 32–33.

Romero-Alvarez, D., Peterson, A. T., Salzer, J. S., Pittiglio, C., Shadomy, S., Traxler, R., et al. (2020). Potential distributions of bacillus anthracis and Bacillus cereus biovar anthracis causing anthrax in Africa. *PLoS Neglected Tropical Diseases, 14*(3), e0008131.

Ruttoh, S. K., Omondi, D. O., & Wanyama, N. I. (2012). Tunga penetrans: A silent setback to development in Kenya. *Journal of Environmental Science and Engineering, 1*(2012), 527–534.

Sentongo, E., & Wabinga, H. (2014). Tungiasis presenting as a soft tissue oral lesion. *BMC Oral Health, 14*(1), 1–4. https://doi.org/10.1186/1472-6831-14-112

Soberón, J. (2005). Interpretation of models of fundamental ecological niches and species' distributional areas. *Biodiversity Informatics, 2*(2005), 1–10. https://doi.org/10.1093/wber/lhm022

Tadele, H. (2017). Clinical profile and outcome of pediatrics tetanus: The experience of a tertiary hospital in Ethiopia. *Ethiopian Journal of Health Sciences, 27*(5), 559. https://doi.org/10.4314/ejhs.v27i5.14

Wafula, S. T., Ssemugabo, C., Namuhani, N., Musoke, D., Ssempebwa, J., & Halage, A. A. (2016). Prevalence and risk factors associated with tungiasis in Mayuge district, eastern Uganda. *The Pan African Medical Journal, 24*, 77. https://doi.org/10.11604/pamj.2016.24.77.8916

Walker, S. L., Lebas, E., De Sario, V., Deyasso, Z., Doni, S. N., Marks, M., et al. (2017). The prevalence and association with health-related quality of life of tungiasis and scabies in schoolchildren in Southern Ethiopia. *PLoS Neglected Tropical Diseases, 11*(8), 1–11. https://doi.org/10.1371/journal.pntd.0005808

Waruguru, C., Mwaniki, P., Karama, M., & Muthami, L. (2015). Prevalence of Tungiasis and its associated factors among residents of Kipkelion west Sub-County; Kericho County, Kenya. *International Journal of Health Sciences & Research, 5*(8), 434–445.

Wiese, S. (2020). *Untersuchungen zu Risikofaktoren für die Präsenz und Intensität der Tungiasis (Sandflohkrankheit) sowie zur krankheitsbezogenen Beeinträchtigung der Lebensqualität von Kindern in Kilifi County, Kenia* (Doctoral dissertation).

Wiese, S., Feldmeier, H., Elson, L., & Larson, P. S. (2015). Household-related risk factors of tungiasis and severe disease in Kilifi County, Kenya. *Tropical Medicine & International Health, 20*, 114–115.

Williams, R. A. J., & Peterson, A. T. (2009). Ecology and geography of avian influenza (HPAI H5N1) transmission in the Middle East and northeastern Africa. *International Journal of Health Geographics., 8*, 47. https://doi.org/10.1186/1476-072X-8-47

Yosef, T., Reszke, R., Melaku, M., & Szepietowski, J. (2018). Disseminated tungiasis in an 8-year-old girl. *Dermatology Review/Przegląd Dermatologiczny, 105*(1), 81–86.

Zabron, W. (2017). *Tungiasis risk factors in rural community in Murang'a county*. Ir-library.ku.ac.ke.

Part V
Conclusion

Chapter 13
A New Era for the Health Geographies in (and of) Sub-Saharan Africa

Jenna Dixon and Isaac Luginaah

Geography's scholarship has engaged with health issues in sub-Saharan African (SSA) for quite some time. As a prime example, John M. Hunter, who is notable for his role in demarcating medical geography as its own sub-discipline, spent much of his career researching diseases in the African Savanah region. His work in West Africa on river blindness as the result of onchocerciasis (Hunter, 1966) was directly responsible for the World Health Organization's 20-year larviciding program to control the disease and "permit economic rehabilitation" (Hunter, 1981). And like others of his time, permeating this work were themes of demographic determinants of health (i.e., a focus on population and ecological capacity) (e.g., Hunter, 1967) and the relationship between disease and development (e.g., Hughes & Hunter, 1970).

Much has changed since then – both in the scholarship of medical/health geography and in SSA itself. Sub-disciplinary debates brought forth new ways of understanding health and well-being, new theoretical engagement, and methodical innovations. In the years that followed, medical and health geographers continued to engage with health issues on the continent, with their focus changing to reflect sub-disciplinary evolutions, such as using critical lenses to expose how power relations create vulnerabilities (Kalipeni, 2000; Kalipeni & Oppong, 1998), or post-structural approaches to expose the impacts of social identities and cultural framings of disease in these experiences (Craddock, 2000). Substantive topics likewise evolved, reflecting the pressing needs of the continent, with much of this scholarship confronting the HIV/AIDS epidemic as it reached critical status within many SSA

J. Dixon (✉)
Faculty of Health and Social Development, University of British Columbia, Kelowna, Canada
e-mail: jenna.dixon@ubc.ca

I. Luginaah
Department of Geography and Environment, Western University, London, Canada
e-mail: iluginaa@uwo.ca

J. A. Braimah et al. (eds.), *Health Geography in Sub-Saharan Africa*, Global Perspectives on Health Geography, https://doi.org/10.1007/978-3-031-37565-1_13

countries (e.g., King, 2010). While these contributions remain seminal to health and medical geography and pushed the sub-disciple forward on a number of fronts, they remain just that – part of health and medical geography. A limited number of scholars doing great work focused on SSA, one part of the larger sub-discipline.

In the past decade or so, the volume of geographic scholarship into health and well-being in SSA has reached a critical mass. It is set against this backdrop that we read and place this edited collection, understanding it as the symbolic turn to a new era of the nascent but stand-alone field of study, what *Braimah, Bisung, and Kuuire* introduce as *health geographies in sub-Saharan Africa (SSA)*. The efforts in these chapters represent a cross-pollination of thinking from two geographical sub-disciplines, health/medical geography and geographies of Africa. It is unsurprising then that many of the authors represented in these pages are found equally in the circles of health geography, evidenced by membership in health geography specialty groups or publications in journals such as *Health and Place*, as they are present and active members of African geography specialty groups, or contributors to journals such as *the African Geographical Review*. As this collection of chapters demonstrates, the scholarship has gone from straddling these two worlds to emboldening a field of its own right – with its own focus, questions, debates, and ethical imperatives.

Running throughout this book is an underlying critique of the simple mental models often used to think about health in SSA, such as the health/development cycle or the demographic transition model. Though they served some use years ago, the work encapsulated here provides a snapshot of a continent reverberating with complexity and spatiotemporal nuances that now defy explanation through those simple models. The complexity is, as *Kuuire, Braimah, Bisung, and Atuoye* (Chap. 2) articulate, attributed to what geographers understand as 'place' – the interweaving of physical (including ecological) and human characteristics, spaces of varying degrees of livability and inequalities/inequities, and people within their contexts. As the continent confronts a triple burden of disease (infectious diseases, non-communicable diseases, globalization, and climate change), the intricacies of place become fundamental in understanding why these simple models fail. Even analytical short-hands for capturing place, such as extrapolating from place-based characteristics, may not be sufficient. To this end, Kuuire et al., while considering non-communal diseases and the role of neighborhoods, offer the readership a vision of three nested layers (individual characteristics, place-based characteristics, and factors contributing to place identity such as historical, political, contemporary global dynamics) as a framework to begin to engage with the complexity at hand in SSA.

Acknowledging all this complexity, it is frustrating to see that the challenges geographers have long been attentive to sadly still remain pressing health problems. On the other hand, it is clear that geographers can offer substantial insight into the dance between environmental and social determinants. Though now half a century after Hunter was working in West Africa, *Deka and Morshed* (Chap. 12) similarly explore linkages between humans and the ecologies in which parasitic diseases flourish. Deka and Morshed focus their analysis in East Africa on Tungiasis disease (as caused by the *Tunga penetrans* flea) which has serious health and social impacts on those who contract it. And while the chapter clearly reaffirms the connection between ecological environments and human health, the authors also make clear that Tungiasis is amplified by the presence of poverty – a human creation.

As critical geographers have well documented, poverty may be experienced locally but it is a multi-scalar creation, and the relational expression of power operationalized on the global stage (Crane et al., 2020). Through this lens, spaces of power throughout the world are implicated in these ecologically bound diseases. Indeed the very existence of neglected tropical diseases (see Chap. 12) are evidence of these unjust power dynamics at play. The neglect for these diseases is not a question of the *ability* to find solutions, but a question of political and economic choices to invest toward it (Aya Pastrana et al., 2020). When people are dying because they are poor, this is an inherently *global* health problem and we must be cautious not to see it as bound to the geographical borders of SSA.

In fact, though this collection does not directly confront some of the historical and contemporary global structures responsible for African exploitation (e.g., European colonialism, globalization), the relevance of the world theatre to the health interests of SSA presents itself as an underlying theme in a number of chapters. This is evident as many chapters frame their contributions in light of the globally created Sustainable Development Goals (see Chaps. 2, 3, and 10) but most exemplary of this is *Smiley, Adams, Agbemor, and Hungerford's* (Chap. 8) case studies on water, sanitation, and hygiene (WASH) interventions during the COVID-19 pandemic from the Democratic Republic of Congo, Ghana, Niger, and Uganda. The pandemic itself was an inherently *global* moment as a virus from abroad threatened those in SSA, but so too were the quick responses outlined in these case studies – seeing the devastation across the oceans, SSA countries were imposing lockdowns and curfews with less than 100 cases on the continent. More so, the chapter makes the case that while there were some successes, WASH-related interventions implemented during the pandemic had many challenges because the system is not set up for quick-fixes. When the global capitalist system commodifies basic necessities such as water, there is a disproportionate impact on populations that face pre-existing vulnerabilities (e.g., migrant and refugee populations), or what Braimah, Bisung, and Kuuire have described in their introduction as structural exposure – which is to say, populations are not vulnerable by accident, but by result of the very ways (global) societal structures are organized.

As the introduction also highlights, the social construct of gender remains a central pillar of analysis as geographers seek to understand the experiences and expressions of health and well-being in SSA. While social and demographic transformations on the continent have brought forth some place-specific changes in the relational experiences of women and men, patriarchy remains a domineering force in women's health decisions. This is a phenomenon explored in detail by *Anfaara et al. in* Chap. 6 who grapple with gendered power dynamics within household settings in post-conflict Democratic Republic of Congo, with a particular focus on postpartum married women. The chapter demonstrates the dominance of patriarchal power systems, as even exposure to media messages about family planning may be determined by a woman's autonomy. The ways that constructs of gender interact with so-called development may also express in multifaceted ways. For instance, *Ezeagbor and Tenkorang* (Chap. 3) consider the non-communicable disease of hypertension, which is most prevalent in women, though often underestimated in part due to gender bias. While non-communicable diseases are associated with 'development' due to changes in lifestyle, the focused look on hypertension in Ghanian women reinforces protective effects of women's formal education on their health outcomes – and thus, again adding nuance to our understanding of health outcomes from these demographic transitions. And in a hopeful turn, *Annan-Aggrey, Kutor, Bandauko, and Arku* (Chap. 10) ask us to consider gender as more than just a lens for analyzing differential health experiences of those in SSA. Instead, they postulate that gender responsive development – that is, the very ways that places are built and run – may contribute to promoting gender equality and securing sustainable livelihoods for all genders. The authors rightly point out that practices that are seemingly 'gender blind' are likely masking discriminatory thinking, as women's priorities are more often than not ignored in designs, provisions, and governance. To counter this, the authors argue for acknowledging women as experts in their own lives and ensuring a wide array of otherwise underrepresented voices are at the table helping to plan cities that work for all.

Though 'health and well-being' are often said as one complete thought, interest from geographers on better capturing concepts of well-being have continued to push scholarship beyond a health-only focus. As a "sense of fulfilment, equilibrium and contentment across multiple aspects of life" (Elliott & Pearce, 2020) Chaps. 4, 9, and 11 contribute to this conversation by considering ways that well-being is experienced in SSA. *Braimah, Dassah, Bisung, and Rosenberg* (Chap. 4) provide a synthesis of existing qualitative research on the coping strategies adopted by persons with mental illness in Ghana. By shining a spotlight onto the range of informal strategies used to overcome the stress associated with this particular illness, they in turn also provide critique toward the lack of support for well-being through the biomedically oriented healthcare services. Drawing on longitudinal data for a participatory farmer-to-farmer agroecology intervention implemented in northern and central Malawi, *Kansanga et al.* (Chap. 9) show a positive relationship between agroecology and self-rated health. While the intervention may have more obvious 'health' advantages, such as offering improved nutrition, the fostering of social interaction and the development of community ties are viewed as essential components for building emotional and mental well-being. And *Kangmennaang, Kuyateh, and Luginaah* (Chap. 11) consider the psychosocial impacts of epidemics though an analysis of (non-)participation in social life after the Ebola outbreak in Sierra Leone. Their findings expose many of the ways viral illnesses may harm well-being, far beyond the usual considerations of the acute crisis. In all three examples, the authors move beyond a focus on the prevalence of illness to contemplate broader ideas of well-being in *interaction* with health and illness in SSA.

Kearns and Neuwelt once cautioned geographers to be careful when doing research from a distance, with a particular concern that the gaze from the Global North onto the Global South has a dangerous tendency to document spaces of deficit, 'needy' populations, and be historical rather than future facing (Kearns & Neuwelt, 2015). This and other parallel conversations have become very important in global health research, with appropriate questions surrounding the positionality of the researcher, how power is exercised, and for whom this research is beneficial (Plamondon & Bisung, 2019). It is worth pausing here to note that the authorship within these pages by and large reflects those born, raised, and/or currently residing in SSA. But issues of positionality and belonging are quite tricky and poorly captured through outsider/insider dichotomies. The intricacies of this are explored by *Tuyisenge, Kasine, and Nisingizwe* (Chap. 7), who contemplate a navigation of their own positionalities doing research at 'home' (in this case, Rwanda) after having been based out of the Global North for some time. The authors unpack simultaneous but conflicting identities as both insiders and outsiders, a task made even more complex by the belief of some that the researcher may exude financial or social power to the benefit of the participant. They offer no standardized solution here but instead remind us (and this is a message that should resonate for all geographers) that reflexivity and critical reflection on our personal 'positionality blind spots' are essential to ethically navigating all research situations.

It is on this note of positionality that we return to reflect on the place of *health geography in SSA*. What marks health geography in SSA as a stand-alone field of study? Does it really matter if the scholarship focused on health in SSA is subsumed under the larger umbrellas of the sub-disciples? While not to say scholars will not remain active contributors to the health/medical geographies and geographies of Africa sub-disciplines, we think that there is indeed value in carving out this space as a field in its own right. As each chapter here has demonstrated, the people and places throughout the continent are multifaceted, the experiences of health are numerous and complex, and, as researchers, the very ways we think about health and well-being needs to adapt to these emerging realities. Geographers have an important role to play in understanding this complexity, but scholars need to engage in conversation with each other about the next steps forward. Recognizing efforts should be taken so that the great insights provided by geographers are mobilized into policy action, and equally mindful of the deep connection contributors in this field have with SSA, perhaps the future here is a health geography *of* SSA as much as it is *in* SSA. We look forward to seeing what comes next.

References

Aya Pastrana, N., Beran, D., Somerville, C., Heller, O., Correia, J. C., & Suggs, L. S. (2020). The process of building the priority of neglected tropical diseases: A global policy analysis. *PLoS Neglected Tropical Diseases, 14*(8), e0008498. https://doi.org/10.1371/journal.pntd.0008498

Craddock, S. (2000). Disease, social identity, and risk: Rethinking the geography of AIDS. *Transactions of the Institute of British Geographers, 25*(2), 153–168. https://doi.org/10.1111/j.0020-2754.2000.00153.x

Crane, A., Elwood, S., & Lawson, V. (2020). Re-politicising poverty: Relational re-conceptualisations of impoverishment. *Antipode, 52*(2), 339–351. https://doi.org/10.1111/anti.12603

Elliott, S. J., & Pearce, J. (2020). Welcome to wellbeing, space and society: A new interdisciplinary journal. *Wellbeing, Space and Society, 1*, 100025. https://doi.org/10.1016/j.wss.2020.100025

Hughes, C. C., & Hunter, J. M. (1970). Disease and "development" in Africa. *Social Science & Medicine (1967), 3*(4), 443–493. https://doi.org/10.1016/0037-7856(70)90022-3

Hunter, J. M. (1966). River blindness in Nangodi, northern Ghana: A hypothesis of cyclical advance and retreat. *Geographical Review, 56*(3), 398. https://doi.org/10.2307/212464

Hunter, J. M. (1967). Population pressure in a part of the west African Savanna: A study of Nangodi, Northeast Ghana. *Annals of the Association of American Geographers, 57*(1), 101–114. https://doi.org/10.1111/j.1467-8306.1967.tb00593.x

Hunter, J. M. (1981). Progress and concerns in the World Health Organization onchocerciasis control program in West Africa. *Social Science & Medicine. Part D: Medical Geography, 15*(2), 261–275. https://doi.org/10.1016/0160-8002(81)90002-2

Kalipeni, E. (2000). Health and disease in southern Africa: A comparative and vulnerability perspective. *Social Science & Medicine, 50*(7–8), 965–983. https://doi.org/10.1016/S0277-9536(99)00348-2

Kalipeni, E., & Oppong, J. (1998). The refugee crisis in Africa and implications for health and disease: A political ecology approach. *Social Science & Medicine, 46*(12), 1637–1653. https://doi.org/10.1016/S0277-9536(97)10129-0

Kearns, R., & Neuwelt, P. (2015). Conclusion: Healthy development/developing health. In R. Bezner Kerr & I. Luginaah (Eds.), *Geographies of health and development* (pp. 297–301). Routledge.

King, B. (2010). Political ecologies of health. *Progress in Human Geography, 34*(1), 38–55. https://doi.org/10.1177/0309132509338642

Plamondon, K. M., & Bisung, E. (2019). The CCGHR principles for Global Health research: Centering equity in research, knowledge translation, and practice. *Social Science & Medicine, 239*, 112530. https://doi.org/10.1016/j.socscimed.2019.112530

Index

J. A. Braimah et al. (eds.), *Health Geography in Sub-Saharan Africa*, Global Perspectives on Health Geography, https://doi.org/10.1007/978-3-031-37565-1